Natural Disaster Reduction

South East Asian Realities, Risk Perception and Global Strategies

ANTHEM South Asian Studies
Series Editor: Crispin Bates

Other titles in the series

Samaddar, Ranabir (ed.) *Women Building Peace Between India and Pakistan* (2007)
Bandyopadhyaya, Jayantanuja *Class and Religion in Ancient India* (2007)
Mukhopadhyay, Surajit C *Conceptualizing Indian Policing in Post-colonial India* (2007)
Fraser, Bashabi (ed.), *Bengal Partition Stories: An Unclosed Chapter* (2006)
CSIRD *Towards BIMSTEC-Japan: Comprehensive Economic Cooperation* (2006)
Banerjee, Sashanka S, *India's Security Dilemmas: Pakistan and Bangladesh* (2006)
Chattopadhyaya, Braj *Studying Early India* (2006)
Sen, Satadru *Colonial Childhoods* (2005)
Bates, Crispin and Basu, Shubho (eds.) *Rethinking Indian Political Institutions* (2005)
Brosius, Christiane *Empowering Visions* (2005)
Mills, Jim (ed.) *Subaltern Sports: Politics and Sport in South Asia* (2005)

NATURAL DISASTER REDUCTION

South East Asian Realities, Risk Perception and Global Strategies

Dilip Kumar Sinha

Anthem Press

ANTHEM PRESS
An imprint of Wimbledon Publishing Company
www.anthempress.com

This edition first published in UK and USA 2007 by
ANTHEM PRESS
75-76 Blackfriars Road, London SE1 8HA, UK
or PO Box 9779, London SW19 7ZG, UK
and
244 Madison Ave. #116, New York, NY 10016, USA

British Library Cataloguing in Publication Data
A catalogue record for this book is available from the British Library.

Library of Congress Cataloguing in Publication Data
A catalogue record for this book has been requested.

1 3 5 7 9 10 8 6 4 2

ISBN-10 1 84331 705 2 (pbk);
ISBN-13 978 1 84331 705 0 (pbk);

ISBN-10 1 84331 704 4 (hbk);
ISBN-13 978 1 84331 704 3 (hbk)

Printed in India

Contents

Acronyms and Abbreviations

AAI	Action Aid International
ADPC	Asian Disaster Preparedness Centre
AID	Association for India's Development
ASEAN	Association of South East Asian Nations
ASI	Archeological Survey of India
BDAT	Barangay Disaster Action Team
CDPC	Central Disaster Prevention Council
CRZ	Coastal Regulatory Zones
DART	Deep Ocean Assessment and Reporting of Tsunamis
DMU	Disaster Management Unit
DOST	Department of Science and Technology
ECMRWF	European Centre for Medium Range Weather Forecast
ED	Explosive Development
ENSO	El Nino Southern Oscillations
ESCAP	Economic and Social Commission for Asia and the Pacific
FAP	Food Action Plan
FRP	Fibre-reinforced plastic
GEF	Global Environmental Facility
GEOSS	Global Earth Observation System of Systems
GHG	Greenhouse Gases
GIS	Geographical Information System
GSN	Global Seismographic Network
HEWS	Humanitarian Early Warning Service
hpa	hectare Pascal
HRFFC	Han River Flood Forecasting Centre
ICDPP	Integrated Community Disaster Planning Programme
ICG	Intergovernmental Coordination Group
ICT	Information and Communication Technology
IDC	International Data Centre
IDNDR	International Decade on Natural Disaster Reduction
IEWP	International Early Warning Programme
IIOE	International Indian Ocean Expeditions
IMD	India Meteorological Department
IMS	International Monitoring System
IOC	Intergovernmental Oceanographic Commission
IOTWS	Indian Ocean Tsunami Warning System
IPCC	Intergovernmental Panel on Climate Change
IT	Information Technology

ITES	IT Enabled Services
JMA	Japan Meteorological Agency
KMA	Korean Meteorological Administration
LDPC	Local Disaster Prevention Council
MAPS	Madras Atomic Power Station
MDG	Millenium Development Goals
MOST	Method of Splitting Tsunami
MW	MegaWatt
NCMRWF	National Centre for Medium Range Weather Forecasting
NOAA	National Oceanic and Atmospheric Administration
NWFP	North West Frontier Province
NWP	Numerical Weather Prediction
ONGC	Oil and Natural Gas Commission
PAR	Philippine Area of Responsibility
PDR	People's Democratic Republic
PFBR	Prototype Fast Breeder Reactor
PHIVOLCS	Philippine Institute of Volcanology and Seismology
PNRC	Philippines National Red Cross
PTWC	Pacific Tsunami Warning Centre
SAARC	South Asian Association for Regional Cooperation
SADMC	South Asian Disaster Management Centre
SEA-EAT	South East Asia Earthquake and Tsunami
SECTRUM	Special Experiment Concerning Typhoon Recurvature and Unusual Movement
SPVS	Short Period Vertical Seismograph
SRO	Seismic Research Observatory
SSARR	Streamflow Synthesis and Reservoir Regulation
SST	Sea Surface Temperature
STC	Scientific and Technical Committee
TC	Tropical cyclone
UNCRD	United Nations Centre for Regional Development
UNDP	United Nations Development Programme
UNEP	United Nations Environment Programme
UNESCO	United Nations Educational, Scientific and Cultural Organization
UNV	UN Volunteers
USDNDR	United States Decade for Natural Disaster Reduction
USGS	United States Geological Survey
WB	World Bank
WCDR	World Conference on Disaster Reduction
WFP	World Food Programme
WHO	World Health Organization
WMO	World Meteorological Organization
WWSS	Worldwide Standard Seismograph
WWW	World Weather Watch
YRCC	Yellow River Conservancy Commission
YRWMP	Yellow River Water Management Programme

Preface

There is an enormous quantity of literature on all kinds of natural disasters. Most of these publications are regularly updated; the accumulation of current data aids projections and future analyses. Because of the unique characteristics of natural hazards at different geographical locations, these studies are often country-based and sometimes, region-based. For example, disasters in Latin American countries differ in many respects from those in the developed world. In developed countries there are more survivors and the loss of property is much lower than in less developed countries. The developing world and the South East Asian region in particular, have suffered far greater devastation from natural disasters, perhaps because of their geography and topography. The accounts of losses in these regions over centuries are appalling. Despite global deliberations on such disasters and the convergence of aid, quite evidently the definition of 'compassionate support' has changed. This was disturbingly apparent in the post-tsunami global relief efforts – strangely, donor countries seemed almost to be *competing* against each other. Gestures that should have been purely humanitarian assumed a rather political hue, probably because of geopolitical compulsions. The efficient flow of funds is the most obvious aspect of disaster relief, but there are other equally important needs. More often than not, survivors of natural disasters desperately need emotional and psychological rather than material relief. A survivor whose entire family has perished cannot be consoled by the petty dole of a free meal or a blanket. Unfortunately, aid has degenerated into a unidirectional flow, with *donors* deciding and dictating what the recipients need and should have.

The 1990s, declared the International Decade on Natural Disaster Reduction (IDNDR), has undoubtedly facilitated a transformation of attitudes towards natural disasters. Nonetheless, the IDNDR and its role have often been scornfully dismissed. The study of natural disasters generally overlooks the cultural diversity of the settings in which they occur. An underlying purpose of this book is to demonstrate that measures for disaster reduction are never independent of a region's culture. However, dialogue fails if concerns about natural hazards and the need to actively help the affected parties do not intersect. Their symbiotic relationship must be recognized. This was one of the fundamental tenets of the IDNDR.

Although natural disaster management is a global concern, the South East Asian region has become the focus of attention after the Indian Ocean tsunami on 26 December 2004. The tsunami and the continuing threat of 'seismicity' and 'tsunamicity' have exposed coastal vulnerability. The Hyogo Declaration at Kobe, Japan in January 2005 was a direct outcome of the World Conference on Disaster Reduction (WCDR). It emphasized the susceptibility of the South East Asian region to natural disasters and asserted that disasters in the region had certain unique characteristics. Local weaknesses, the responses and aspirations of communities in disaster-prone areas are markedly different from those elsewhere. Yet common concerns must be globally recognized if the directives of the IDNDR and WCDR are to be followed. Synergy between global support and the ground realities of natural disasters is urgently needed. This is the *raison d'etre* of the book.

The following chapters do not merely gloss over the causes of these disasters but investigate the entire range of challenges presented by them. The book attempts to dovetail the effects of the recent tsunami and a résumé of global perspectives presented at the IDNDR and WCDR and the conclusions of regional meets held under the aegis of the Economic and Social Commission for Asia and the Pacific (ESCAP), the United Nations Centre for Regional Development (UNCRD) under the United Nations Development Programme (UNDP), the World Meteorological Organization (WMO), the United Nations Educational, Scientific and Cultural Organization (UNESCO), etc. based in South East Asian countries. The post-disaster response of South East Asian communites has been separately analysed. Besides a broad discussion of natural disasters, the recent tsunami has been examined in detail.

The contemporary emphasis on sustainable development, poverty reduction and the competitive nature of corporate forces should expand the book's readership. Countries such as Japan, China, Indonesia, Vietnam, North and South Korea, Thailand, Philippines, Bangladesh, Myanmar, Maldives, Malaysia, Singapore, Taiwan, Cambodia, Nepal, Pakistan, Sri Lanka and India are in the process of revaluing their policies on and management of natural disasters. This study should be relevant for members of educational institutions, government organizations, NGOs and other agencies. The development of new methods of sensitization and capacity building measures has ensured that a considerably large sector is involved directly or indirectly with natural disaster management.

I would like to thank Anthem Press for offering me the opportunity to write *Natural Disaster Reduction* for their *South Asian Studies* series. I am particularly grateful to my editor, Anirban Sarma, for his enthusiasm and his numerous suggestions for the book's improvement. I thank Mr B Chatterjee, an amanuensis par excellence. A work like this cannot be written without technical advice from experts. I am indebted to Mr D P Sinha and Dr D Sinha for their help and guidance. I also thank the members of my family for bearing with me as I undertook this rather onerous task. I look forward to inputs from readers and critics - the value of a book is best assessed by the lively debate it generates.

DILIP KUMAR SINHA

Kolkata
26 December 2006
Two years since the tsunami.

Prologue

Any commentary on natural disasters must refer to the International Decade on Natural Disaster Reduction. Indeed, its acronym, IDNDR, has passed into popular parlance. The fairly well recorded deliberations on natural disasters and associated activities and programmes in the last few decades of the twentieth century need to be treated as precursors to certain phenomena in the twenty-first century. Unlike the Agenda 21 Declaration that was formulated in Rio de Janeiro in 1992, there was no equivalent international agreement in the area of disaster/ tsunami control when the 26 December 2004 tsunami ravaged the coastlines of the Indian Ocean.

The fact that Indonesia, Thailand, Myanmar, Sri Lanka, Bangladesh and India (countries adjoining the Indian Ocean) and even distant Somalia were caught haplessly unaware despite adhering to the tenets of the IDNDR, exposes their extreme vulnerability. Tsunamis and their consequences have been analysed at length; thus it is deplorable that the affected region did not have an early warning system like the one in the Pacific Ocean that alerts neighbouring countries to underwater seismic activity. The tsunami that hit South Asia on 26 December was a catastrophe of such extraordinary dimensions that it arrested the attention of the whole world and generated new debates about natural disasters, their consequences and preventive measures.

Statistical details of the South East Asian tragedy are now available in abundance. The death toll, details about damage to/ loss of property and facts and figures pertaining to aid are easily accessible. Human responses, roles and responsibilities form a triad that has assumed high priority. The comprehensive Hyogo Declaration in January 2005 was a prompt reaction to the tsunami. The IDNDR did not have a similar precedent. Although the IDNDR's efficiency and ability to act promptly has been questioned, it certainly monitored global trends, facilitated the international sharing of knowledge and the progress of science and technology. It also focused on relief efforts to mitigate the after-effects of disasters.

Apart from the 2004 tsunami itself and the recent World Conference on Disaster Reduction (WCDR), the Asian summit at Jakarta in 2005 followed by the ministerial meeting at Phuket, Thailand greatly stimulated debates and discussions about natural disaster management. The Asian Disaster Preparedness Centre (ADPC) at Bangkok studies disasters in Asia; its findings are based on the population distribution map representing 500,000 people affected by natural hazards. In terms of population density, the variety and frequency of natural hazards, Asia is far more susceptible to disaster risks than any other continent. The reports of the ESCAP also testify to the vulnerability of the South East Asian region.

Although the IDNDR was launched with much fanfare, many were skeptical whether it would be able to keep pace with changes in the latter part of the decade. It is not enough to measure the success of efforts in a single decade. Care must be taken to ensure that 'development' does not paradoxically lead disaster management astray. The South East Asian region has been working assiduously to expand the scope and scale of national initiatives in the field of disaster management. The poser remains: 'What sort of spadework is necessary to build national capabilities in South East Asia?' A host of meetings, lectures and reports have been prepared and organized by the ESCAP, but it is a moot question whether Asian countries have something tangible to show for their efforts. Of the many positive efforts during the IDNDR, the provision of support to proactive countries that have evinced their success needs to be prioritized. The IDNDR exercised its functions with a certain degree of enthusiasm and zeal and it remains to be seen whether the WCDR can instill a similar attitude in natural disaster activists. The WCDR must try to make South East Asia—without neglecting other regions—the hub of worldwide disaster management practices. The global recognition of the region as a disaster-prone zone goes in its favour. Lately, disaster management has become technically very complex. Emergency management is combined with inputs from communication and information technology (IT) and the well-known Geographical Information System (GIS). Area management has to deal with intrinsically non-linear elements. Even in the post-disaster phase, one has to rely heavily upon the creativity and entrepreneurial efforts of the affected community.

The father of the IDNDR, Prof. Frank Press recognized the interdependence of individuals, communities, nations and the economic linkages between them. He perceived that natural disasters are 'shared event[s] with repercussions far outside the immediate area of impact'.[1] Indeed, these far-reaching 'repercussions' are demonstrated by the mathematics of Indian assistance to affected developing countries like Sri Lanka, Maldives, Thailand, Malaysia and Indonesia. The apparent contraction of geographical distance as a result of globalization amplifies the repercussion of natural disasters. James Gustave Speth points out the need for a 'profound commitment to urgent action'.[2] This kind of concerted action must be undertaken by academics, activists, policymakers and corporate honchos. Partnerships and collaborations are absolutely essential for countering the devastation caused by natural hazards, e.g. relief activities for the 2004 tsunami were conducted under the aegis of the 'Framework Partnership Agreement'. As Peter Berger, the Techinical Associate of the European Commissions Humanitarian Aid Office (ECHO) remarked: 'Of course the capacity and quality of work of agencies is continuously evaluated. We can monitor through spot checks but also work closely with NGOs to build a positive partnership.'[3]

The shared risk of communities vulnerable to disasters and the dire threat to citizens, private and public organizations has hitherto been inadequately addressed. The notion of 'risk' is rapidly becoming more and more complex—hence risk management needs to be accorded far greater importance on the policy agenda. 'Global risk' has established itself as a contemporary buzzword, and South Asia in particular must undertake a variety of urgent enterprises in the context of 'shared disaster events'.

The agenda for contingency/crisis management must be globalized. Cutting-edge advances in science and technology are indispensable to disaster management. However, incorporating global trends into local milieus is an uphill task. The nature and aspirations of South East Asian communities, brought to the fore by the recent tsunami, need to be woven into the fabric of natural disaster management. This book assimilates the thoughts and ideas of South East Asian communities into its multi-faceted study of natural disasters in the region.

NOTES

1. 'UNDRO News, January/February 1988, pp.12-3, 26-7.
2. Speth, James Gustave, *Global Environmental Challenges: Transitions to a Sustainable World*, Orient Longman, New Delhi, 2004.
3. *The Sunday Times*, Kolkata, 9 January 2006

Chapter 1

Overview of Disaster Scenarios in South East Asia

INTRODUCTION

A fairly comprehensive overview of natural disasters across the world may be constructed with information available on the World Wide Web. For the purposes of this book, it is useful to classify natural disasters according to type and the countries of their occurrence. A geographical outline of each country in South East Asia (excluding the ones in the Pacific) enhances one's understanding of natural disasters *in situ.* The ESCAP is a leading organization working in the field of natural disaster management and there are several others, like the UNCRD. In the following pages, natural disasters in South East Asia have been discussed in order to distill their essential characteristics.

I. CHINA

Tropical cyclones cause more casualties, injuries and losses than any other kind of natural hazard in China. About one-third of the world's tropical cyclones (TC) occur in the western North Pacific basin. TCs here occur with an intensity greater than or equal to that of storms. The majority of tropical storms assume the form of typhoons or super-typhoons. Figures show that while the countries along the western North Pacific Ocean suffer the most severe damages because of typhoons, China has the most severe typhoon disasters in the world. One could refer to the committee reports on typhoons by ESCAP/WMO. Obviously, one should expect timely typhoon warnings and fairly accurate track forecasts, particularly where one often finds sudden changes of typhoon tracks and intensity and torrential rains following typhoons. Hence there are intensive efforts in China to improve the typhoon track forecasting capability. Close observation has revealed that topographical features such as islands, coastal lines and mountains have a bearing on the occurrence of typhoons. For example, the island of Taiwan is responsible for a low pressure over the Taiwan Strait when a typhoon approaches. Surprisingly the original typhoon slows down and dissipates quickly while a newly induced low develops, replacing the original typhoon with another having greater acceleration. Coastal lines and mountains also influence the direction and velocity of typhoons. The inner structure of a TC has an impact on typhoon motion. A better understanding of the physics of TCs, particularly by considering their asymmetric structure, is necessary for more accurate track forecasting. Observations indicate that typhoons tend to avoid the cold sea areas and approach the warmer regions.

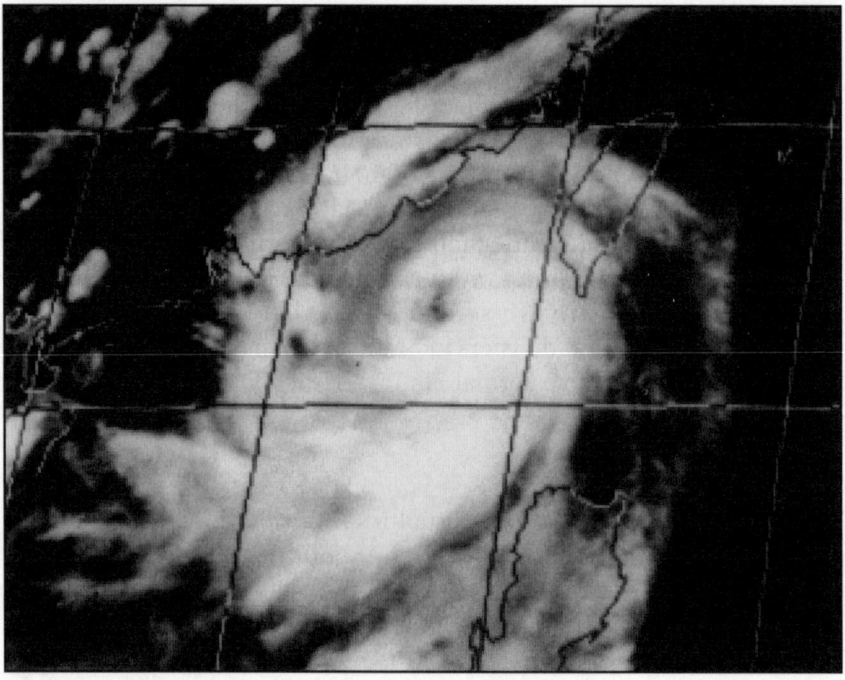

Fig.1 GMS-4 infra-red imagery at 1200 UTC on 28 June 1990 as the eye of Typhoon Percy (9006) reformed over the South China Sea.

Disaster reduction in China is considered an important strategy for developing the Chinese economy; thus efforts are being made to reduce disasters caused by typhoons. Natural disasters in China are caused mostly by heavy rain, strong winds and storm surges. The government and the ruling party attach great importance to public awareness about natural disasters; policies evolve from their persistent efforts to create awareness with the help of the State Meteorological and Oceanic System. Numerous instances of death and suffering in the 1920s and 1930s may be cited, but over the decades the losses have been reduced even though the intensity of typhoons is the same or even greater. There is a continuing effort to improve disaster reduction using advances in science and technology. Several Doppler radars, weather radars and satellite signal receiving stations with modern in-built communication systems are being used, mostly in the eastern coastal regions. A striking feature of such ventures is the use of typhoon history databases, situational databases, economic, demographic and geographical databases. The IDNDR has directed the Chinese efforts to develop a powerful and capable National System of Disaster Reduction in China with an associated State Information System for Natural Disasters, a central communication net, a State watch and warning system, a computer system, a State centre system, etc.

It has been observed that El Nino Southern Oscillations (ENSO) influence the variability of tropical storm formations over the Indian Ocean. There is a relationship between storm activity and ENSO, which in turn depends on seasonal changes. The explosive development (ED) of typhoons is an important area of research. By ED we usually mean a decrease in pressure of 40 hpa or more in the centre of a typhoon within 24 hours. This development is considered important because it is often responsible for wrong predictions that cause heavy loss to life and property in the coastal regions.

On an average the South China Sea and Taiwan Strait are hit by TCs up to 160 days a year. These TCs often create waves that exceed a height of 4 metres along the continental shelf close to the coast. Such raging waves claim hundreds of lives and cause innumerable shipwrecks. Hence, there is an urgent need to investigate the evolution of typhoon-induced waves. Problems of wind-sea evolution during typhoons assume crucial importance. A typhoon wave protection system must necessarily include sea surface pressure and wind data: the development of such a system is being undertaken because of the damage to coastal structures, the losses to agriculture, etc. As disasters caused by typhoon waves are among the principal marine disasters in China, great importance is attached to the direction of typhoon waves as part of the work on sea wave forecasts.

A 'storm surge' is a marine phenomenon characterized by the abnormal rise and fall of the sea level that results from a drop or rise in air pressure and the strong winds of a volatile weather system. Storm surges affect the coastline of China which is more than 1,800 km long with about 6,000 islands and a wide continental shelf whose eastern and southern low-lying areas are easily flooded. China is frequently visited by storm surges mainly caused by TCs. Chinese oceanographers and meteorologists use monitoring and forecasting techniques and conduct theoretical studies of storm surges chiefly because of the complex system of continental shelves with many semi-closed bays, estuaries, canals and islands in China's seas. Surface wind stress is a major factor inducing storm surges along the coast of China. For a fixed region or station, the TC track is usually the key factor determining the surge magnitude. There are more than 200 tidal gauge stations in China's coastal provinces for storm surge monitoring and warning: they can transmit warnings about tidal waves by telegraph on a real-time basis to the State Oceanic Administration whenever a storm approaches China. A typhoon storm surge in 1952 drowned around 100,000 people in the mouth of Zhu-Jiang river and the strong surge generated by Typhoon No.7 in 1980 raised the water level to such a great height that the vast fertile lands in Zhangiang and Haikou were flooded, physically injuring about 1,000 people. Hence disasters caused by typhoon surges are taken very seriously by the Chinese authorities.

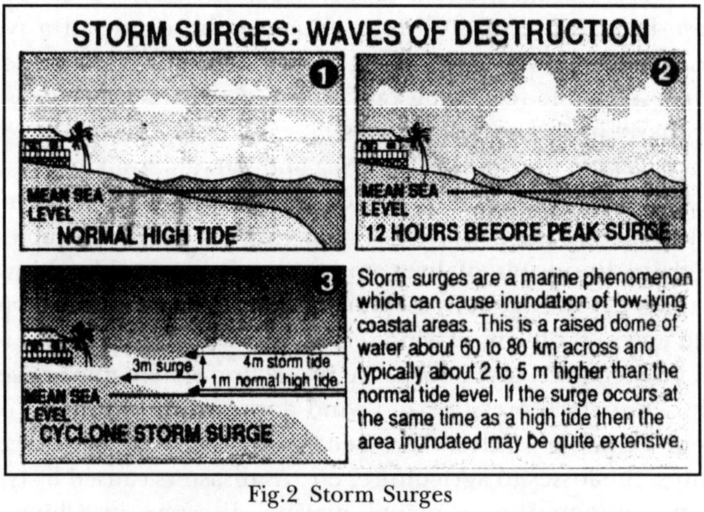

Fig.2 Storm Surges

TCs on land in China cause torrential rains. Their formation can be attributed to a variety of factors such as the maintenance of the TC circulation system, stagnation of TCs, steady supply of water vapour, interaction between the circulation systems of low latitudes and those of middle latitude, influences of mesoscale systems, osographic systems, etc. The TC is a kind of quasi-symmetrical weather system in which isobars are circular in form and the associated area of gales (of a certain scale) is roughly circular in shape, particularly in the open sea.

After landfall, the wind forces reduce rapidly and the areas of strong winds also reduce accordingly. The occurrence of heavy rainfall diminishes rapidly with the reduction of TC intensity after landfall but is often intensified and prolonged by the interaction of other weather systems. Although torrential rains due to TCs usually occur after the landfall of the TC centre, heavy rainfall occurs when the centres of relatively strong typhoons move close to the mainland. The precipitation from the near-shore TCs generally being less than that of TCs over land, few of them can induce heavy rainfall. Much can be said about the relation between TCs, their intensities, their directions of motion, duration of rainfall, etc. In some cases there is almost no intense precipitation within the TC circulation area but at its periphery, where heavy rainfall occurs. Such TCs that mature at a late stage are called 'hollow typhoons'. TCs occurring with significant precipitation but having strong winds are called 'dry typhoons'. Thus one can construct a typology of TCs accompanying torrential rain. Torrential rain itself may cause flash floods and mudslides.

Torrential rains occur not only along the Chinese coasts but also in the hinterland of central China. The decay of typhoons over land and its corresponding relationship with rainfall in areas away from the interior poses problems for investigation by meteorologists trying to correlate their findings on landfall with exceptional rainfall. From the scientific standpoint, it is often held that the low level of frictional dissipation and upper level of horizontal export of kinetic energy comprise the major energy slumps for typhoons, while the torrential rain area chiefly has a kinetic energy source which can be attributed to the formation of the rain and the release of latent heat. This, of course, leads to the evolution of the kinetic energy export and thus the rapid dissipation of the tempest or typhoon can be explained. South China, the Hunan and Hubei basin, the eastern coastal area and the Huainhe valley are the areas where, on an average, flooding occurs once or twice in two or three years. There are also records of flooding and waterlogging in the Yellow River and Chang Jiang river valley. China experiences frequent hailstorms, but since these occur more on mountains than on plains, inland and coastal areas, they do not cause major losses.

China has three climatic zones: tropical, extra-tropical and cold (from south to north). East China faces the Pacific Ocean and the Qinghai-Tibet plateau, known as the roof of the world, is located in west China. The complex geography of China makes it susceptible to monsoons, changeable weather and variable continental climate. The Chinese traditionally refer to TCs as typhoons. The Chinese have suffered from typhoons formed in warm seas having a vortex with a low atmospheric pressure at the centre. There are around seven TCs with maximum wind speeds of over 17.2 m/s in China each year and an annual average of 20 TCs in sea areas. Typhoons occur in China from May to December but they occur in the adjacent seas nearly all year long. Most parts of China are exposed to typhoon hazards. The strong winds, heavy rains and storm surges in China's coastal areas have been mentioned earlier. Consequently, China's inland areas also experience heavy rainfall. For example Typhoon Nina hit the coast of Fujian province on 4 August 1975, moved inland for two days, was converted to a tropical depression and became more or less steady on 7 August, but it caused 1,605 mm of rainfall from 5-8 August at Linz Huang, Henan province. This resulted in a devastating flood, destroying reservoirs and railways and ravaging the Beijing-Guang Zhow province. Typhoon Sally struck the coast of south China in September 1996 and moved on to cause further damage in Vietnam. Sally is often chosen as a representative instance of typhoons in this region. Sally entered the South China Sea emanating from the Philippines, travelled rapidly towards the coast of Guangdong and moved towards Hong Kong. It was a full-fledged typhoon with a maximum wind speed of 49 m/s at its centre while moving towards the coast of western Guangdong. With a constant speed of 10 m/s over the northern part of the South China Sea, it is considered an extraordinarily fast moving typhoon. Wind speed gained rapidly in the late evening with squally showers and thunderstorms as it sped toward Hong Kong. The direction of the wind then changed from north-easterly to south-easterly, as it moved about 200 km to the south at around 2 am on 9 September. It continued toward the port of Zhan-Jiang in the Guangi province, and finally moved into north Vietnam. It caused very serious damage, particularly in western Guangdong and Guanji provinces; the cities worst affected were Zhan-Jiang, Maoming and Yang Jing where patterns of life were totally disrupted and there was widespread loss of lives and property. Although the loss of lives in each of these areas did not exceed a three-digit figure due to meteorological forecasting using satellite data in China, it appears that the warnings about Typhoon Sally went unheeded. In response to the frequent recurrence of typhoons, the Shanghai Typhoon Institute has been set up to study various aspects of marine disasters and surge dynamics from theoretical and practical standpoints. Often typhoons are said to possess 'swarmability', meaning that several typhoons may strike the same region during a short period. For example, three tropical storms struck most parts of north-east China almost simultaneously from 2-25 August 1985 causing severe flooding and waterlogging. There have been many similar incidents. It is mainly because of these events that the systems of typhoon monitoring, forecasting and warning are being assiduously enforced and acted upon. Some of these steps have already been indicated.

Over the years, China has responded proactively to the bilateral and global strategies on disaster reduction. Chinese scientists are inclined towards strategies for the ready flow of information. The initial lack of communication during the 1976 earthquake at Tangshan, resulted in terrible losses and caused the deaths of about 650,000 people.

Seismic disaster reduction and its countermeasures in China have engaged Chinese scientists, particularly those belonging to the Seismological Bureau of Shaudong province, Jinan and the Institute of Geology, State Seismological Bureau, Beijing. Earthquake prediction is undertaken on the basis of seismic disaster analyses. Based on prior knowledge of risk areas, long-term prediction and monitoring tasks are being pursued to improve social resistance and prevent seismic disaster in China. However, it has been observed that earthquake predictions are not entirely reliable and their social value is not taken seriously enough. Therefore, necessary emergency measures have been taken up by the government to reduce seismic disasters. Chinese scientists under the aegis of the Institute of Atmospheric Physics, Chinese Academy of Sciences, Beijing are involved in theoretical and experimental forecasts of droughts in north China. Their predictions have been fairly satisfactory.

The physical details of the two rivers Yangtze and Hwang Ho (Yellow River) are well known. There is a vast amount of statistical data about their large discharge, gentle slopes, enormous sediment deposits and the large population in their floodplains. The socio-economic conditions of the surrounding areas are heavily dependent on these rivers. Consequently, floods in these rivers always cause overwhelming damage in the surrounding areas.

The Hwang Ho (Yellow River) has the greatest sediment content among the major rivers of the world. The silt deposition has affected the riverbed and the risk of heavy flooding prevails. The Yellow River Conservancy Commission (YRCC) has made some trenchant observations about the river's changing characteristics. The Commission points out that extreme rainfall along with melting snow and ice cause flooding of the river, but the maintenance of the levee system has declined due to the high growth of population in the North China Plain. Lately, the massive levee system and adoption of flood control steps have improved the situation considerably. The banks of the river have been enlarged and innovative techniques of construction have been used. The Yellow River Water Management Programme (YRWMP) has done much to aid the flood control work. The YRCC has tried to reduce flood hazards through the construction of large dams and reclamation projects. People still identify China's sorrow with the Hwang Ho. The YRCC in collaboration with the World Bank (WB) has developed a Yellow River Economic Model, whose objective is to reduce flood loss.

The Yangtze is the longest river in China and the third longest in the world. It witnessed flood damages of the worst kind in the first year of the IDNDR, largely due to heavy rainfall in eastern China. The Chinese government mitigated the disaster by building dykes, draining flood waters and of course, delivering relief assistance. The People's Liberation Army and China Armed Police were requisitioned to deal with the devastation.

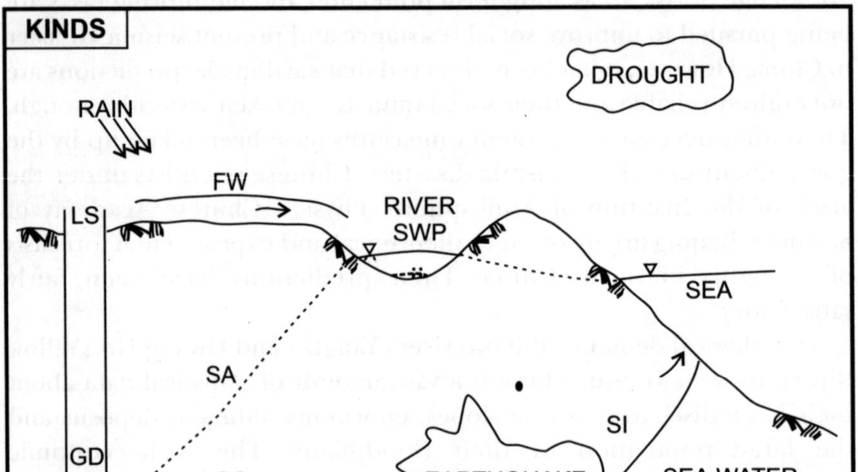

FW — Flood & Water Logging; SI — Seawater Intrusion;
SA — Salinization & Alkalinization; S — Sedimentation;
LGQ — Low Groundwater Quality; GD — Groundwater Drop;
LS — Land Subsidence; SWP — Surface Water Pollution

Fig.3 The kinds of disasters in the North China Plain

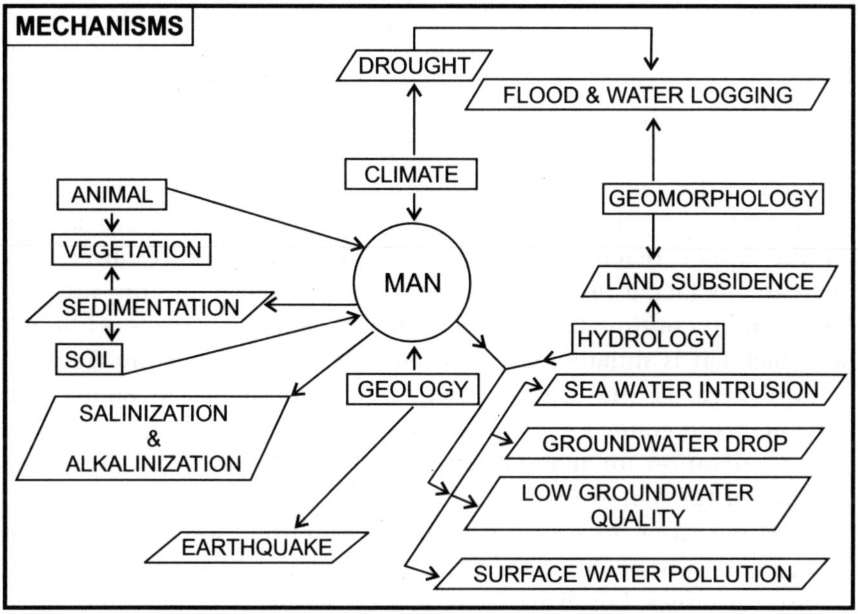

Fig.4 The mechanisms of disasters in the North China Plain

In order to control flooding in the Yangtze basin, a new project has been taken up under the aegis of the Global Environmental Facility (GEF) whose implementing agencies are the United Nations Development Programme (UNDP), United Nations Environment Programme (UNEP) and the WB. Initially a grant of US $4 million has been arranged, to be supplemented by a further grant of US $23 million to support China's efforts to reduce floods in the Yangtze basin, which houses one-third of the country's population. In 1998 catastrophic floods killed 1,562 people, affected another 2.29 million and caused a direct economic loss of about US $20 billion. The scale and frequency of Yangtze floods over the years demand sustainable development of the threatened basin. Hence, with support from the GEF, the Government of China is applying integrated eco-system management techniques to at least two sites to control floods in the wider interest of the upkeep of the sites' biodiversity. Perhaps because of GEF support, the sequestration and reduction of carbon emissions are also part of the charter of programmes. Another spin-off of the support from the UNEP via GEF is to establish an early warning system to improve the response to future floods. This is being done in close collaboration with China's State Environmental Protection Administration. The Chinese government feels that if the plan is successfully implemented for the entire Yangtze river basin, it may be replicated elsewhere.

In June-August 1998, there were disastrous floods in the Yangtze valley. The area of the valley is about 1,800,000 sq km and it forms about 18.8 per cent of the land area of China. The floods caused direct losses of 300,000 million yuan (US $36,000 million). They are attributed to environmental changes caused by the loss of natural vegetation and soil erosion in the valley's upper reaches, affecting 393,000 sq km (more than one-third of the total area). Drought also causes enormous damage in the upper reaches which though rich in water resources are increasingly facing problems of water scarcity and pollution, with the river run-off continually decreasing due to climatic changes. Although west Sichuan is situated at a high altitude and suffers from frequent natural hazards, the use of science and technology has enabled its inhabitants to combat such disasters. Many programmes and activities are in operation for maintaining the socio-economic and ecological balance of the Yangtze river valley.

In Hong Kong heavy rain is caused by either summer monsoon troughs or typhoons that are often accompanied by landslides, floods and storm surges. Low-lying areas near the coast are prone to flooding. Flash flooding occurs in small catchment areas due to intense rainfall. Floods in Hong Kong often pose serious problems because of a critical combination of all the phenomena mentioned earlier. The Chinese relief efforts in response to recent flood damages, particularly the Yangtze floods, demonstrate their efficiency and competence. This is often attributed to the strong centralized decision-making system of the country as a whole.

Perhaps the most recent typhoon in the region struck the eastern parts of China around 22 July 2005. Whether the phenomenon in Zhejiang province was a typhoon or strictly speaking a tropical storm is debatable, but it affected 7.64 million people according to media reports, destroying 7,553 houses and severely damaging more than 58,000 others. The cost of damage was reportedly about US $870 million. Several infrastructural facilities (relating to conservancy), river dykes, hydrological stations, traffic, power supply and telecommunications were badly hit.

Typhoon Haitong's tracks were recorded in Taiwan before it landed on the south-eastern coast of the Chinese mainland on the afternoon of Tuesday, 19 July 2005. The previous evening, the eastern coastal areas of Zhejiang had experienced unusually strong winds accompanied by storms. Most parts of Zhejiang experienced heavy downpours from Sunday to Wednesday. In particular, there were torrential rains over Wenzhou, Taizhou and Lishui. Four counties of Wenzhou reportedly received rainfall of over 500 mm and the Cangnan county reported the highest rainfall of 699 mm. As Typhoon Haitong later dissipated into a tropical low pressure, some parts of Zhejiang province were left relatively less damaged. The province reported that the storm affected 500,000 acres of crop, of which over 90,000 acres were totally destroyed. Crops and aquatic products were also adversely affected.

The provinces of China which suffer most from typhoons are Zhejiang, Fujian and Guangdong; similarly droughts occur mainly in the three north-eastern provinces of Heilonggiang, Jilin and Liaoning.

Typhoon Talion ravaged the Chinese provinces of Zhejiang, Anhei and Zianxi, affecting around 2 million people. The Anhei province had a death toll of 39 caused by a heavy downpour followed by floods and landslides. According to a preliminary report the Dabie mountain area (having a population of around 1.80 million people in 12 counties) was most severely affected. Infrastructure, cropland and houses were destroyed. The estimated loss was about 7 million yuan (US $86.3 million). The rainfall in the upper reaches of the Xianghongian reservoir located near the Dabie mountain exceeded 400 mm in 24 hours, a rare phenomenon in contemporary history. The water level in 47 of the 112 large- and medium-size reservoirs in the province was found above the danger line. The Zhejiang province experienced thunderstorms and landslides caused by the typhoon and at least 15 people were killed. About nine others were reportedly missing in Wenzhou city (in Zhejiang province).

II. JAPAN

Japan has been affected by earthquakes more than any other kind of natural disaster. Earthquakes in turn have often caused tsunamis and landslides. On many occasions seismic activity has wiped out lives, property and infrastructure. According to Gilbert White about 10 per cent of the world's earthquakes occur in and around the Japanese islands. The most destructive earthquakes have occurred in the outer zone of the Japanese sea, off the Pacific Coast. The Hokkaido-Sanrikeu coast and Tokaido-Nankaido coasts still appear to be active seismic zones, recording tremors of more than 8.0 on the Richter scale although the magnitude of earthquakes occurring near the Japanese sea coasts seldom exceeds 8.0. The Nobi earthquake in 1891, however, was an exception with a magnitude of 8.4. The Tokai and Kanto areas are the most active earthquake zones. Other areas experiencing earthquakes are the western Akita prefecture, north-western Yamagata prefecture, eastern Miyagi prefecture, eastern Fukushima prefecture, south-western Nigata prefecture, north Nagans prefecture, western Nagano prefecture, eastern Gifa prefecture, the Nagoya-Kyoto-Osaka-Kofe area, the eastern Shimane prefecture and the vicinity of Iyo-Nada and Hyuga Nada. Some of the areas mentioned in the latter group, have experienced large-scale earthquakes in the past but not recently, while some of them are still in the tectonic area. Some others have shown recent crustal movement and of course, some of them such as Tokyo, Nagoya and Osaka are of critical social importance.

It has been observed that large-scale earthquakes have recurred in the region of the boundary of subduction of the Pacific Plate and the Philippine Sea Plate beneath which the Eurasian Plate is situated. Since more than 130 years have passed since the occurrence of the last major earthquake in the area of Suruga Bay, the possibility of another large-scale earthquake cannot be ruled out. In fact, a Tokai earthquake may occur in the near future.

Any large-scale earthquake causes a variety of changes and inflicts serious damages. As we know, violent tsunamis strike coastal areas after being generated by undersea earthquakes. One could cite numerous instances of earthquakes in the aforementioned regions with their return periods. The Nigata earthquake of 1964 like the Kanto earthquake of 1923 motivated a building code, particularly in the nearby urban areas. Since the Nigata earthquake caused tsunamis and even fire, originating in big petroleum tanks, the impact was devastating. Most of the areas submerged by floods caused by the tsunami were well below the mean sea level because of land subsidence caused by withdrawal of groundwater. The Tokachi-Oki earthquake of 1960 partially destroyed the city of Tohoku. These disasters prompted Japanese seismologists to probe deeper for the causes of earthquakes. The seismologist, Kioo Wadati, made a breakthrough as a young employee of the Japan Meteorological Agency (JMA). He amassed relevant data and applied new methods to determine earthquake depth. His research on the frequency and intensity of earthquakes in Japan enabled the Japanese government to establish the world's best network of seismographic stations. Much earlier, in 1935, Wadati had published a map showing the focal depth of earthquake sites near Japan. Wadati, along with another seismologist Hugo Benioff, wrote a seminal work on earthquake patterns. The study of the science of earthquakes, known as seismology, has an edge over other methods of studying natural disasters. On an average, Japan experiences one destructive earthquake every three years. Thus the Japanese have pursued the study of natural hazards very seriously indeed. The emergency operations of the Earthquake Research Institute swing into action immediately after the occurrence of an earthquake. There are various other organizations that record aftershocks, study the geology of the region, locate and map faults and investigate the structural reasons for the collapse of buildings. As reported by Don Leet, governments tend to opt for extensive and precise surveys to fix the horizontal position and the level (or the vertical position) of closely spaced networks on the islands. An important outcome of this was the discovery that the earth's crust is composed of a mosaic of irregular blocks. Don Leet points out that the mobility of the earth's crust is the basic and immediate cause of earthquakes.[1]

Fig.5 1964 earthquake, Nigata, Japan

Of all the large-scale earthquakes, the Great Kanto Earthquake of 1923 deserves a special mention. The Japanese became especially cautious after this event and opted (at least in theory) to build earthquake-resistant cities. Nevertheless, the Great Hanshin Earthquake that occurred on 17 January 1995 caused widespread damage. The latter earthquake was an eye-opener for the international community. The collapse of old wooden houses (100,000 buildings were completely destroyed and around 285,000 partially destroyed) was the primary cause of the 5,500 fatalities. A parallel may be drawn with the Latur earthquake in India in which people were crushed to death while still asleep, when their houses collapsed. A series of horrifying events followed the Latur earthquake. 215 fires broke out on the same day burning about 85 hectares of the city. To the global community, this recalled the Michoacan earthquake in Mexico City and the Loma Prieta earthquake in San Francisco.

The areas most affected by Great Hanshin Earthquake included the city of Kobe (located in the southern part of the Hyogo prefecture), the northern part of Awaji island and several parts of the eastern Osaka prefecture. Ironically enough, Kobe was the venue of the WCDR held on 18-22 January 2005 that resulted in the Hyogo Declaration.

The comprehensive study of the Great Hanshin Earthquake by the United Nations Centre for Regional Development (UNCRD), Nagoya, Japan in 1995 merits close examination. This report deals not only with micro-seismicity regarding the origin of the earthquake but also discusses several tectonic implications of the earthquake. The earthquake measured 7.2 on the Richter scale, with the epicentre located about 15 km to the south-west of Kobe. The report provides useful information about the disaster prevention system in Japan. In 1961 the Disaster Countermeasures Basic Act was formulated and the Government of Japan formed the Central Disaster Prevention Council (CDPC) on the basis of this Act. Obviously there are hierarchical establishments, but each administrative sector in Japan has instituted its own Local Disaster Prevention Council (LDPC) on the basis of the Act. The emergency measures and responses are also taken care of by these bodies in collaboration with various divisions and subdivisions. The Hyogo prefecture suffered the worst damage due to the Great Hanshin Earthquake as the district lay just above the earthquake's point of origination. Relief and rehabilitation operations were carried out immediately by the Kobe city authorities; they also ensured search and rescue activities with the participation of the local community. Members of non-government agencies and volunteers rescued people trapped under collapsed buildings and distributed supplies and medicines to victims. A striking feature of the report is that it dwells on the economic and psychological effects of natural hazards and the role of foreigners in grappling with this event. The conclusion of the report is particularly revealing in its situation of disaster management in the context of a disaster of such magnitude. The final section of the report deals with necessary countermeasures; expert committees investigate the possibility of future earthquakes and point out how existing countermeasures can be improved.

Water-related natural disasters, particularly typhoons, ravage Japan at least once a year. The Meteorological Agency assumes responsibility for the various phases of typhoons, particularly the pre-disaster phase. Through the 1940s and 1950s Japan experienced strong typhoons which caused widespread damage. Around 1,000 people were killed or went missing every year, and the annual death toll often exceeded 4,000. In 1959 Japan witnessed a monstrous typhoon along the coast of the Ise Bay. Associated with a storm surge of 3.5 metres observed at Nagoya, the Ise Bay typhoon claimed 5,000 lives. Many coastal leaves were broken during the inland motion of the typhoon. The coast of the Ise Bay suffered land subsidence before the typhoon hit the shore. The inundation lasted over a month, covering an area of around 300 sq km. One can conclude that the storm surges in Japan are mostly caused by typhoons and Tokyo, Nagoya and Osaka are highly vulnerable to storm surges. The Ise Bay typhoon eventually led to the formation of a National Research Institute for Earth Science and Disaster Prevention and the typhoon research department in the JMA. These events took place in 1960, a year after the Ise Bay tragedy, and they were followed in 1961 by the promulgation of the Act mentioned earlier. Storm surges in the Bay of Bengal, the east coast of the United States, the coast of the Gulf of Mexico, the north-east coast of Australia and the coast of East Asia indicate that large storm surges occur along the coasts of vast shallow seas. In reality, bays in Japan have vast shallow sea areas, and thus there exist long fetch lengths of strong southerly winds to generate storm surges. This is well reflected by the formula $h = CFW^2/D$, where 'h' is the sea level rise, 'F','W' and 'D' are the fetch length, wind speed, water depth and a numerical constant respectively.

The Japanese establishment, after the Ise Bay typhoon, studied paths of storm surges which helped prepare an inundation hazard map. A striking feature of damage to properties located parallel and perpendicular to the direction of the storm surge is that the damage in the latter direction is more than that in the former.

JMA has developed methods for forecasting storm surges, particularly the method seeking to describe the motion of the seawater, the atmospheric pressure and wind. Indeed, such attempts were reinforced after the terrible Ise Bay typhoon so that the simulation technique developed could be applied to other major bays.

Japanese ports facing the European sea experience storm surges, like the one generated by the typhoon at Tosa Bay leading to a peak surge of 2.1 metres at Kochi port. There are several very active groups that monitor and forecast storm surges. Japanese scientists have identified some structural measures for storm surge prevention and these consist of levee banks, floodgates locked over the land and drainage pumps and facilities which are looked after by the local metropolitan government or even by the port management body, the bay being at Tokyo. Japan defends itself against typhoons with the help of embankments, revetments and wind break forests. Breakwaters, revetments, jetties and detached breakwaters are built in defense against tsunamis and storm surges.

Typhoons mainly occur in August-October. Extra-tropical cyclones occur primarily in northern Japan and Nagasaki prefecture in Kyushu district; heavy rains caused by frontal storms occur in the northern part of Kyushu district, the Kii peninsula and Diphu prefecture. Besides the Ise Bay typhoon in 1959, two other violent typhoons in the areas surrounding Nagoya may be mentioned: the 1934 typhoon in and around Osaka and the Muroto typhoon near Muoto Cape in 1961.

The powerful Typhoon Nabi moved across Japan on 6 September 2005 with a speed of 126 kmph. Heading towards the Sea of Japan, it swept over Kyushu in a matter of hours. It moved north at 25 kmph swamping the archipelago with more than 1,000 mm of rain. The typhoon brought with it violent winds of 90 kmph or more across a radius of nearly 300 km, reportedly wider than the 220 km Hurricane Katrina covered at its peak. Typhoon Nabi was, however, weaker than Katrina. According to initial reports, around three to four people died or went missing in the southern part of Japan, mostly in landslides. The Kyodo News Agencies reported that one woman died in the western Okayama prefecture by falling off a ferry struck by waves, while a man was killed in the Tokyo area. Rapid measures were taken to evacuate and shelter people likely to be affected, because the weather agency predicted that the typhoon would unleash strong winds over a wide area. Besides, the horrifying images of destruction caused by Katrina were sill fresh in everybody's minds. More than 100,000 people were evacuated from their homes in Kyushu. The combination of a heavy downpour and the slow speed of the typhoon exacerbated the situation and kept the government on tenterhooks. There were harrowing accounts of the effects of consequent landslides. The typhoon also brought heavy rains in other parts of the country for which a small number of fatalities and injuries were reported. An estimated 21 people were killed.

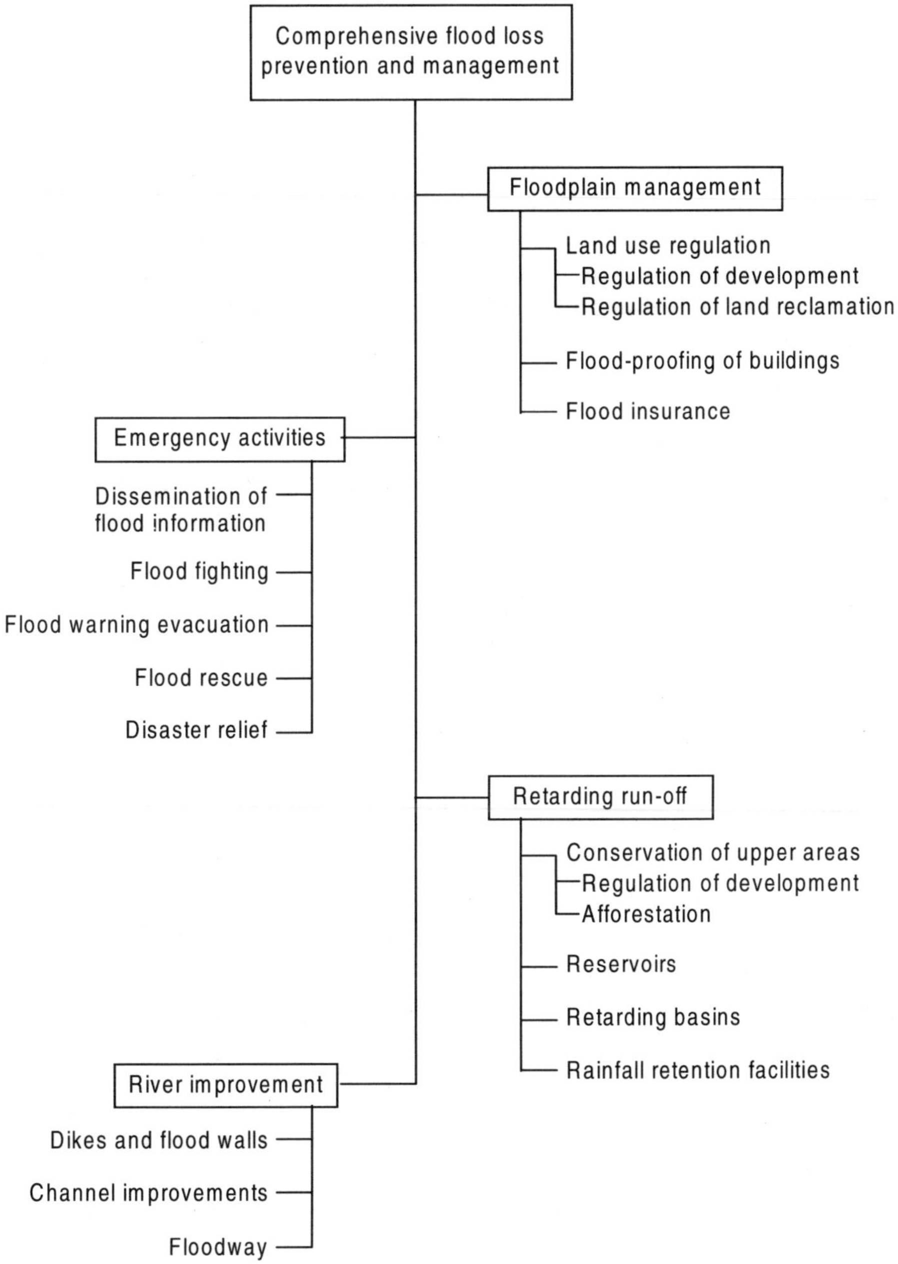

Fig.6 Comprhensive strategy for flood loss prevention in Japan

Of the volcanoes in the Japanese islands, around 80 are still active. Volcanic eruptions in earlier centuries resulted in lava flows, mud flows, blasts, pyroelastic flows, gases, landslides, earthquakes and tsunamis. The worst eruption was that of Bandaisan, Tonoku in 1888. Several lakes were formed in the valleys due to the filling up of rivers by mud flows, several hectares of forest and farm land were destroyed and there were deaths in many villages. Thus volcanic activity not only causes natural disasters but also creates new landscapes. The Taishoike lake at Kami Kochi in the Japanese Alps was formed by the eruption of Yakedake in 1915. One must also mention the eruption of Sakurajema, Kullshu district in 1914. Although some of the inhabitants fled to the mainland and some could be evacuated, around eight villages were buried and several people killed by lava flows and ashfall. It is observed that volcanic activity occurs in periodic cycles. The initial years of the twentieth century witnessed regular eruptions. For instance, Oshima an island in Tokyo Bay, Asama about 250 km inland north-west of Tokyo and Sakurajima experienced violent eruptions. Similar eruptions recurred in Asama (1908), in Osima (1912) and in Sakurajima (1914). The Sakurajima eruption was especially horrifying. At 10 am on Monday 12 January 1914, there was an explosion on the west side of the mountain and ten minutes later the east side blew out releasing ashes which rose to enormous heights. One-twelfth of the entire mountain had been blown to pieces. Volcanoes often cause alarming drops in temperature; for instance following the Asama eruption in Japan in 1973, the average temperature dropped well below the annual mean of any year between 1750 and the 1930s.

In the first week of July 2005 an underwater volcanic eruption reportedly caused a column of water vapour to shoot 1 km into the sky from the Pacific Ocean near the small island of Iwo Jima. According to the JMA, there did not appear to be any danger of a tsunami caused by undersea seismic activity. The massive column of dense clouds was visible about 50 km south-east of Iwo Jima, about 1,120 km south-east of Tokyo. Fairly recently there was a similar occurrence at the same location. The ocean surface in the area where the column rose looked red, indicating underwater volcanic activity. In order to minimize the losses caused by active volcanoes, Japan has undertaken several measures under the Disaster Countermeasures Basic Act, such as the improvement of refuge facilities, disaster resistant farming facilities, volcanic ash control facilities, etc. Besides dealing with conventional disasters, the Act also includes countermeasures against snowfall, fire due to hazardous materials and maritime disasters.

A tsunami is generally caused by an undersea earthquake. It is thus often described as a seismic sea wave, but it can also originate from volcanic eruptions. The Pacific coast is struck by sea waves originating near the Hokkaido and Rywkui islands. Even the tsunamis originating in the Chilean coast of South America cause damages to the Pacific coast of Japan—a classic example of this occurred in 1960. The Sanrikeu coast and the eastern coast of the Kii peninsula are constantly threatened by tsunamis and undersea earthquakes. The repetition of violent tsunamis in the region highlights the susceptibility of the Pacific coast to such natural disasters. The Meiji Sanriju tsunami of 1896 caused by far the worst damage. It originated with an earthquake measuring 7.6 on the Richter scale in the offshore area of the Sanrikeu coast and hit the regions between Hokkaido and the Oga peninsula. A major determinant of the power of a tsunami is the wave height to which its energy converges, and here it was more than 24 metres. 27,122 were killed and 9,316 injured; 10,617 houses and 7,032 ships were destroyed. Most tsunamis in the coastal area of Japan, e.g. the one at Hokkaido-Nankaido, have been caused by earthquakes. These have killed thousands of people and destroyed houses and ships. Worse damages have been recorded along the coastlines of the Izu and Kii peninsulas and the southern coast of the Shikoku island. The Tokyo and Osaka bays are also prone to tsunamis.

The Chilean tsunami reached the Japanese coast around 22-4 hours after the earthquake and the maximum wave height exceeded 6 metres in some parts of the Sanrikeu coast. Losses were great—even valuable resources like pearl beds were destroyed by violent currents at Shima, located on the eastern coast of the Kii peninsula.

Although the coastal inhabitants of Chile could not be warned of the tsunami itself, warnings were issued about 10-30 minutes after the earthquake. Various measures are taken to predict earthquakes, essential for preventing tsunami damages. The Pan Pacific tsunami warning system has been very helpful in matters of detection, prediction and warning. Apart from earthquakes and volcanoes, tsunamis can also be generated by submarine landslides; in these cases they are not accompanied by the 'foreshocks' which are common in case of earthquakes and volcanic eruptions. A submarine landslide is less effective than an earthquake as a tsunami source. In the case of an earthquake, the ground motion is a vertical disturbance but for landslides, ground motion is horizontal. The National Research Institute for Earth Science and Disaster Prevention is studying the efficiency of generation of tsunamis caused by landslides. Specific tsunamis are studied at universities, e.g. the Nihonkai-Chubu earthquake and the tsunami of 1983 have been examined thoroughly. The time of arrival, the tsunami's propagation and the run-up heights of the first and second waves have been studied in detail by scientists at the Disaster Control Research Centre, Tohuku University, Japan.

III. INDONESIA

The Indonesian archipelago is composed of five main islands and about 30 smaller groups, totaling around 13,677 islands and islets of which 6,000 are fairly densely populated. The latter group is situated in the tropical zone between the Pacific and Indian Oceans; it thus bridges the continents of Asia and Australia. The geographical and geological position of the Indonesian archipelago exposes it to a variety of natural disasters. The Republic of Indonesia stretches from 6°8' North latitude to 11°15' South latitude and from 94°45' to 141°65' East longitude, with an estimated area of about 5,193,250 sq km, of which the land area is 202,787 sq km and sea area 3,166,163 sq km. Java, with a surface area of 132,187 sq km, is the most fertile and densely populated area. The island of Borneo comprises Kalimantan, Sulawesi and Irian Jaya. The other islands are smaller in terms of their geographical area and population size.

Over the centuries Indonesia has suffered terrible seismic activity with an annual average of around 350 earthquakes, of which four to ten earthquakes are of particularly great magnitude and occur mostly in the southern island. Figures indicate that approximately half to two-thirds of the total number of South East Asian earthquakes have occurred in Indonesia. In terms of seismological settings, the Indonesian archipelago can be classified into three groups: (a) the Alpine Java or Tethys Mountain System stretching southward from west continental Asia, running west of Sumatra and subsequently turning east south of Java towards the Moluccan islands (b) the Circum-Pacific System forming a part of the East Asian System, extending from the islands of Japan all the way down to the Philippines, the northern part of Sulawesi and the Moluccan islands and (c) the Circum-Australia Belt forming a part of the East Asian System, stretching from New Zealand through Australia, the province of Indian Java (which was hit by two severe earthquakes within three weeks in June-July 1976) and the Moluccan islands.

Indonesia is severely affected by volcanic eruptions. There are about 130 active volcanoes that cause a high rate of earthquakes. The

Indonesian islands often experience active crustal movement. The region stretching from Sumatra, passing through Java and the Lesser Sunda islands to the north and following the western part of Sulawesi, are identified as areas particularly prone to earthquakes and volcanic eruptions. Java has around 35 centres of volcanic activity of which 16 are identified as 'type A' volcanoes. Of the 130 active volcanoes in Indonesia, there are around 75 type A volcanoes. Some of the active volcanoes are particularly destructive. The 1883 Crakatoa (located between Java and Sumatra) volcanic eruption killed about 36,000 people and a decade ago, there were 31 eruptions in Galunggung in west Java. Mt Keludin in east Java erupted on 10 and 12 February 1990 killing 30 people, and around 45,000 people had to be evacuated. As in several other parts of the world, Indonesia's fertile soil is continually replenished by volcanic eruptions, e.g. in the island of Java the ejected lava congeals into enormously fertile soil.

Since Indonesia experiences a large number of volcanic eruptions, earthquakes and various forms of crustal motion, the occurrence of tsunamis cannot be precluded. The 26 December 2004 tsunami was triggered by an underwater earthquake off the west coast of northern Sumatra in Indonesia. The earthquake measured 8.2 on the Richter scale. Strictly speaking, it should be measured in terms of MW, the moment of magnitude which measures the force of an earthquake's total seismic energy released as a function of rock rigidity in the fault, the total area of contact where friction occurred and the amount of slippage (or displacement). The Sumatra earthquake measured 9.0 MW. The ensuing tsunami struck many coastal areas around the northern Indian Ocean and caused loss of lives and other damages. The epicentre of the earthquake in Sumatra was reportedly at latitude 3.9° North, longitude 95.9° East, which was in the Pacific Ocean's Ring of Fire, one of the world's most volatile seismic belts. This earthquake is believed to have been of tectonic origin—the Indian plate and the Myanmar plate came into contact with each other and stress walls built up as one plate pushed down the other. The Myanmar plate was snapped off at a distance of about 1,200 km thereby causing a massive displacement of water in the Indian Ocean, and consequently waves began travelling at 800 kmph in all directions. Tsunami waves, despite their long span, slow down and acquire greater heights when they strike shallow waters near the coast.

The situation in Indonesia was appalling; 25 per cent of the population of a single town, Banda Aceh, were reported dead. The majority of victims were women and children. In fact, more than a third of the victims were children. The energy released by the earthquake was massive and some parts of Sumatra and smaller nearby islands shifted up to 35 metres after the earthquake.

A similar event occurred on 12 December 1992 when a violent tectonic earthquake measuring 7.5 on the Richter scale struck the island of Flores at 13.29 local time, and was immediately followed by a strong tidal wave which hit the northern coast of the district of Sikka, eastern Ende (on Flores Island) and Ngada killing 2,080 persons and injuring 3,103. The islands of Babi and Pomana Besar located in the Flores Sea and inhabited by 2,000 people were severely affected by the tidal wave. Flores with its population of about 700,000 persons is located 1,750 km east of Jakarta. As an anticipatory measure, several thousands of people had to be evacuated because 90 per cent of the town of Maumere was destroyed. The damage estimate was around US $100 billion. Landslides followed by severe rainfall and mudflow occurring in Ndod on 28 December 1992 were also responsible for the earthquake. An earthquake followed by a tsunami rocked the Nusa-tengaria island of eastern Indonesia on 19 August 1977 in which more than 95 people were killed and 20,000 rendered homeless.

On 27-28 May 2006 an earthquake measuring 6.2 on the Richter scale struck Yogyakarta in central Java. Yogyakarta is the cultural capital of Indonesia, 440 km east of Jakarta. The earthquake occurred in the early hours of morning when most people were still asleep, causing greatest damage to Banful. Its epicentre was very close to the volcano Mt Merapi. As a result of the quake Mt Merapi spewed large clouds of hot gas and ash for weeks. One such eruption caused debris to spread along the volcano's western flank for about 3.5 km.

An equally powerful earthquake (measuring 6.3 on the Richter scale) shook eastern Indonesia on 1 December 2006, killing one woman, damaging dozens of buildings and sending terrified residents fleeing to higher ground shouting 'Tsunami!' Striking at 10:01 pm (GMT 14:01), the earthquake's epicentre was 25 km north of Raba, a town on Sumbawa island. Fortunately, the earthquake did not trigger a seismic wave.[2]

Indonesia's location in the Pacific Ring of Fire, an arc of volcanoes and fault lines encircling the Pacific basin, makes it especially susceptible to earthquakes, volcanic eruptions, tsunamis, high tides and other associated natural disasters. In its policy on disaster management, the Government of Indonesia is emphasizing preparedness and mitigation programmes and has established the National Coordinating Board for Disaster Management.

IV. PHILIPPINES

The Philippines is an archipelago consisting of over 7,000 islands. Though it is situated in the south-eastern part of Asia it also features prominently in the equatorial western Circum-Pacific Belt of Fire. Therefore, geographically speaking Philippines stands in the trajectory of turbulent and destructive typhoons of which about 20 occur within the Philippine Area of Responsibility (PAR). The country is ravaged regularly by typhoons, causing enormous damages and loss of lives. The land mass stretches for nearly 1,600 km between the southern tip of Taiwan and the northern part of Borneo and Indonesia. The archipelago has a breadth of 1,120 km and is bounded on the east by the Pacific Ocean and on the west by the South China Sea. Commercially, it is strategically located on the China Sea on the major sea routes for transporting goods and materials from Europe and the Middle East to South East Asia and the Pacific through the Indian Ocean, the Malacca Strait and the China Sea.

The islands of the Philippines form the higher portion of a partly submerged mountain chain and are therefore largely mountainous throughout. Many mountains are part of a girdle of active volcanoes around the Pacific Ocean, often called the Circum-Pacific Volcanic Belt; thus volcanic eruptions and associated earthquakes occur occasionally in the region.

Since the land mass (area 3,000 sq km) of the Philippines comprises mostly plains and valleys, these typhoons flood the terrain heavily. In fact, in around mid-November 1990, Typhoon Mike (Philippine name 'Ruping') destroyed a vast area, the hardest hit places being the Visayas and northern Mindoro. According to the Filipino people, who suffered heavy losses of lives and property, Ruping was the thirteenth tropical cyclone that entered the PAR; it was, in fact, seventh in terms of intensity and destructive power. The winds that accompanied the typhoon had velocities of 130-240 kmph while the gusto ranged from 160-180 kmph. Most of the motor vessels that ran between the islands were lost, fishing boats (Bancas) were washed ashore and even overturned by the violent high waves generated by the typhoon. Over a hundred ships sank because of the typhoon, more than 800 lives were lost and the damage to property was an estimated US $10.8 billion.

Besides typhoons, the Philippines have been adversely affected by volcanic eruptions and earthquakes on numerous occasions. Thus the Philippine Institute of Volcanology and Seismology (PHIVOLCS) was established under the Department of Science and Technology (DOST), Philippines. Two typical cases have been discussed below: one of a volcanic eruption and the other of an earthquake followed by a tsunami in a Philippine island. Both occurred during the IDNDR.

Mt Pinatubo is situated in the central Luzon of the Philippines. It was inactive for 600 years until it erupted on 16 June 1991. Among earlier volcanic eruptions, the one at Mayon in 1814 and Taal in 1911 caused a thousand or more casualties. The Mt Pinatubo eruption killed more than 800 people and over 250,000 people had to be evacuated. In July 1992 there were floods, landslides and mud flows which rose 1.6 metres high destroying bridges, roads and dwellings in the provinces of Pamtanga, Tarlac and Zambales, north of Manila. There was a threat of another eruption in August that year, but instead Mt Pinatubo triggered mud flows and secondary explosions along the Sacobia river. There were continuous torrential rains and the disaster resulted in at least 50 dead, 13 injured and ten missing. The number of affected persons rose to 942,764. In the plains of Luzon, the entire rice and sugar cane crop was swept away by several tons of ashes and rocks carried along by the violent rains during three successive typhoons. Many historical and heritage sites were destroyed or suffered severe damages. The overall damage was estimated at approximately US $74.2 million.

There is a great deal of seismic activity in the scenic island of Mindoro, situated 130 km south of Manila and considered to be the seventh largest island in the Philippines with a population of more than one million. The island has two provinces located east and west of the mountain in the central part of the island. Mindoro has a history of being affected by typhoons; for example, prior to the Mindoro shock in 1994, three successive typhoons occurred between December 1993 and January 1994. A major earthquake, measuring 7.1 on the Richter scale, struck the island on 15 November 1994. It struck at a depth of 12 km, causing 78 human casualties, destroying 7,566 dwellings and affecting 22,452 families. The damages aggregated about 500 million pesos (2.08 billion yen). The Mindoro has a rough terrain with an irregular coastline, particularly Oriental Mindoro or the eastern part of the island. Several mild earthquakes have occurred in northern Mindoro; these quakes were mostly shallow, with the deeper ones occurring in the main island of Luzon. The Mindoro earthquake occurred at 3:15 am local time on 15 November 1994 (GMT 19:15, 14 November) with its epicentre near Verde island (13.5° North, 121.1° East). The UNCRD in collaboration with the PHIVOLCS has undertaken an excellent study of the Mindoro island earthquake that includes a variety of observations on the nature of faults involved, ruptures caused, etc.

At about 4:26 am on 16 July 1990, the Philippines experienced a disastrous earthquake and the country was rocked by a series of tremors felt through most of Luzon and as far as Samar in the Visayas. PHIVOLCS reported an intensity of 8 in Neuva Ecija, Pangasinan, Tarlac, Benguet, La Union and south-west Nueva Vizcaya. An intensity of 7 was felt in Manila, Pampanga, Bataan, north-eastern Nueva Vizcaya, eastern Jambalies, northern Bulacan and south-eastern Quirino. Around 15,000 sq km were heavily affected. It appears that there were two main shocks, the first of which had its centre some 10 km south-east of Cabanatuan City, followed by a stronger tremor with its epicentre at Baguio City.

The earthquake triggered a tsunami that destroyed a number of barangays on the northern shore of Oriental Mindoro and in the islands north of Mindoro. The tsunami was caused by the second M 7.1 earthquake. This was not unusual, for a tsunami is generally caused by a deep-slip earthquake of magnitude greater than M 6.5 and with a focal depth not less than 50 km.[3] The tsunami caused major damages along the shallow coastlines at Baco, Malaylay and Pulong Malaki, chiefly because of the topography of these areas. Much like the tsunami that rocked the Indian coastline, this one too claimed many lives. It was only in Pulong Malaki (Baco Island), where the tsunami run-up wave was countered by a strong run-down, that one child was carried away by the wave and remained missing.

The devastation was distributed over the barangays in Malaylay, old Baco, Brgys, Waya and Pulong Malai (on Baco Island). The earthquake also triggered a minor seiche with a vertical run-up of about 0.5 metres on the northern shore, particularly at Lake Naujan in the north-eastern part of Oriental Mindoro. This caused a minor inundation and the waters of Lake Baruyan overflowed, endangering fishermen and local residents. According to the media, because of the proximity of the earthquake's epicentre and the time of occurrence, no advance warning could be given. Many were drowned and buildings were heavily damaged. Although the people of Mindoro have experienced other natural disasters, the impact of this earthquake and its consequent tsunami was immense. The provincial government and several other agencies, including the defence forces, responded to the urgent need for relief, recovery and rehabilitation. Seismic and tsunami countermeasures were planned for the various stages of such natural hazards.

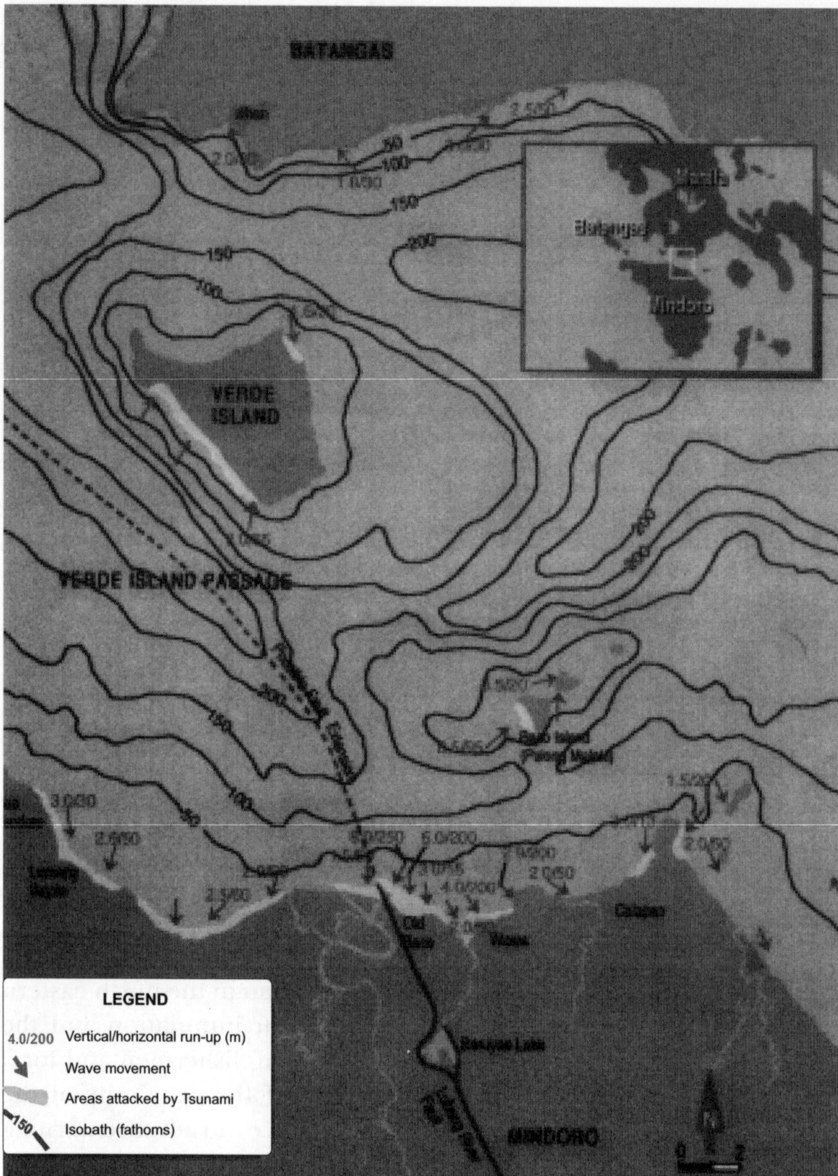

Fig.7 Areas affected by tsunami showing vertical and horizontal run-ups (PHIVOLCS)

As recently as 30 November 2006, Typhoon Durian blasted the Philippines causing massive mudslides that smothered mountainside villages. The death toll exceeded 300 and officials feared that it could

Fig.8 Houses destroyed by tsunami at Baco island (Photo: PHIVOLCS)

double to 600 or more in the days ahead. As of 3 December, 300 people were still missing. The Albay province was the worst hit—the province's Disaster Coordinating Council reported 236 dead, including 129 in the town of Guinobatan swamped by flood waters in the Mayon volcano's foothills. Three towns on Mayon's slopes experienced severe mudslides of volcanic ash and boulders, and each reported at least 22 dead. Houses along the Yawa River in Padang, about 10 km from the capital of Albay province, were buried under five feet of mud, with only their roofs protruding. Rescue operations swung into action promptly, with the Philippines National Red Cross (PNRC) delivering food, tents, water and body bags to the disaster sites.[4]

On the whole, the Philippines has proved itself capable of coping with the adverse effects of frequent natural disasters. Improved forecasting methods and warning systems for all types of disasters have helped minimize the loss of lives and property.

V. MALAYSIA

Malaysia consists of two parts separated by around 700 km of the South China Sea. Peninsular Malaysia forms the southernmost portion of the

South East Asian mainland. The other part is in the northern quarter of the island of Kalimantan (formerly Borneo), comprising the two sparsely populated states of Sabah and Sarawak. Peninsular Malaysia is situated between 01°16' North and 06°43' North and between 99°39' East and 104°118' East. Peninsular Malaysia comprises 11 states, namely Perlis, Kedah, Palau, Penang, Perak, Selangor, Megeri Sambilan, Melaka, Johor, Pahang, Terengganu and Kelantan. Peninsular Malaysia lies just north of the equator; being a narrow peninsula it has a mild equatorial climate and is usually free from seasonal monsoon rain, except for some parts, i.e. states like Kelantan, Terengganu, Pahang and east Johor. Being parts of an island, the states of Sabah and Sarawak between 0°52' and 07°12' North and between 109°32' and 119° 16' East seldom have extreme climate, except, of course, during the annual monsoon season. Situated south of the typhoon-prone zone, Sabah escapes the effect of typhoons. In the local dialect 'Sabah' means 'land below the wind', indicating that it is outside the wind (typhoon) zone. Similarly, though Peninsular Malaysia lies fairly close to the volcanic belt, it is far enough to be safe from volcanic eruptions or other natural disasters like earthquakes. Floods are the only major natural hazard in Malaysia, though the country also experiences droughts and minor landslides. Flash floods occur several times a year in rural and urban areas and especially on agricultural farms. Severe floods occur every 10-20 years causing severe damages. Minor droughts occur once in 5-10 years affecting agricultural lands and minor landslides block roads affecting transportation.

Historically, human settlements in many parts of Malaysia have been situated wholly or partly in floodplains or coastal lands prone to river floods. Over the years, because of the expansion of settlements in flood-prone areas, occurrences of flood hazards have greatly increased.

Perhaps the worst flood of the twentieth century occurred in 1926, causing extensive damage in the Pahan-Terengganu border region. Subsequently there were major floods in 1931, 1947, 1954 and 1957. In 1967, major floods caused heavy losses in the states of Kelantan, Terengganu and Perak. Four years later in January 1971, another major flood occurred in Peninsular Malaysia, affecting eight of its 11 states, including its capital Kuala Lumpur. Malaysia has taken a variety of (structural and non-structural) countermeasures to combat the floods. A computer-based flood forecasting system was installed in 1973 and it was subsequently upgraded with the installation of a telemetric rainfall and river station network. There are extensive disaster preparedness programmes and relief committees at different levels. The National

Disaster Control Centre (also known as the Flood Control Centre) has several tiers—federal, state and district. The dissemination of information on flood situations is undertaken by the Malaysian meteorological services.

Floods in Malaysia are mainly of two types—monsoon floods and flash floods. Monsoon floods are, by and large, influenced by the Asian monsoons, i.e. by the north-east and south-east monsoons. Monsoon floods are thus an annual affair in Malaysia, affecting a large part of the country. Such floods may often last for several weeks before abating, causing widespread damage to crops and property, disrupting the livelihood of the people and even claiming lives. These floods take place both in coastal and inland areas and also affect urban centres, because in Malaysia the rainfall causing such floods is invariably heavy.

Flash floods are a common and frequent phenomenon resulting in disasters of varying degrees. These floods occur without prior warning signals. Malaysia has addressed its flood problems quite seriously in order to reduce the loss of lives and property. Flood mitigation and management programmes depend upon an effective information structure at different levels. A schematic diagram of flood disaster phases and their information requirements is given below:

During the summer season in the northern hemisphere (June to September) rainfall over the west coast of Peninsular Malaysia is augmented by typhoons close to Vietnam. This may be attributed to the lifting of potentially unstable air by the mountain ranges, but a reduction in rainfall is observed when the typhoons are farther away. Typhoons on the east coast of Peninsular Malaysia occur between 10° North and 15° North leading to drier conditions, perhaps because of the suppression of local effects by the strong synoptic wind. In Sarawak, rainfall tends to be subnormal during tropical storms. Moreover, typhoons are likely to reduce rainfall when they are north of 15° North and this is also observed between 10° and 15° North when typhoons move towards the east of Philippines. Sabah is more susceptible to typhoons than any other region in Malaysia and has supernormal rainfall when tropical storms are nearby and subnormal rainfall when such storms are far from it.

If the tropical storm is sufficiently close to its shore, Sabah has supernormal rainfall from October to early November. Following this, i.e. from late November to December (early winter monsoon), states on the east coast of Peninsular Malaysia have spells of heavy rain associated with cold surges. It has been found that storms in the Philippines and South China Sea influence the temperatures on the east coast. Rainfall

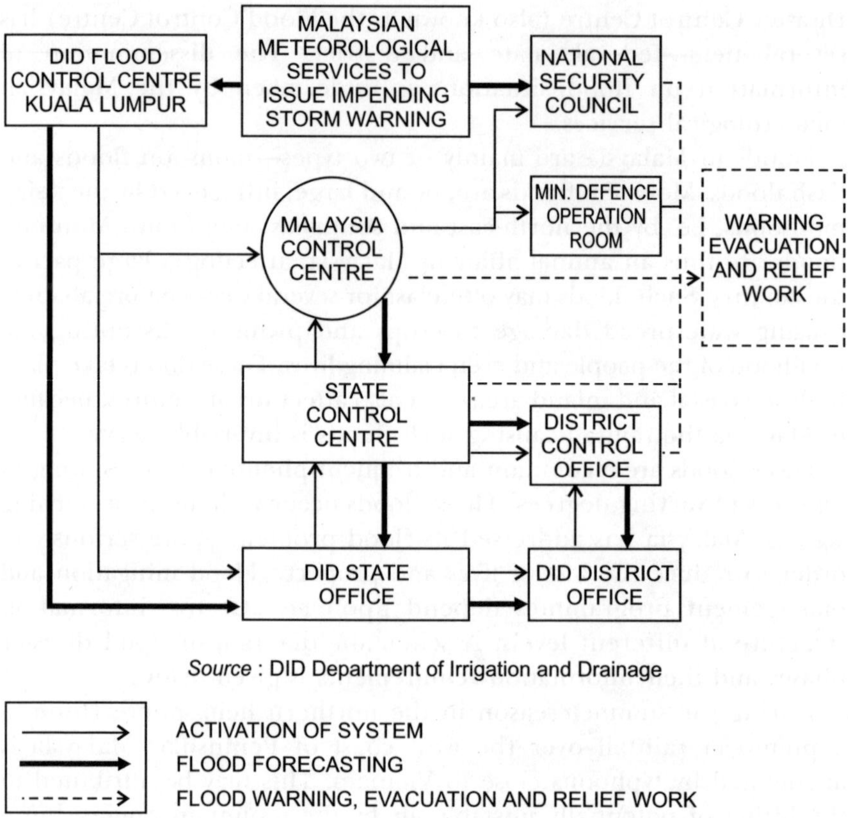

Source : DID Department of Irrigation and Drainage

Fig.9 Flood Forecasting and Warning Practice in Malaysia

in Malaysia is thus somewhat indirectly related to the status and intensity of tropical cyclones by modifying the normal circulation over the Malaysian region (strengthening the circulation charges in the flow direction and shifting the monsoon from its mean position). In terms of physics, such changes in circulation cause fairly significant variations in the Malaysian region in some dynamic and thermodynamic parameters, which eventually lead to fluctuations in rainfall.

It is said that the influence of Typhoon Yancy extended as far as equatorial Malaysia and thereafter an attempt was made to study the large-scale impact of this typhoon on Malaysian weather. This study was brought within the purview of the Special Experiment Concerning Typhoon Recurvature and Unusual Movement (SECTRUM) during the IDNDR. It was observed that Peninsular Malaysia and Sabah recorded convective rainfall amounts before Yancy intensified into the typhoon stage, with a virtually negligible amount of rainfall the next day that coincided with its stage of maturity. Thus rainfall was caused by a typhoon distant from Malaysia but having the potential of convective activities during its west-north-westwards movement towards China.

Malaysia has undertaken studies on flood risk for the upper Klang river basin, 1,200 sq km in area, located in the state of Selangor, Darulrhsan and the federal territory. This basin has a drainage area of 712 sq km and river length of 56 km. The climate of the basin is tropical, subject to monsoons but not typhoons. Daily average temperature ranges from 23° C to 32° C; the annual average rainfall is about 2,450 mm. Kuala Lumpur has been subjected to flooding due to the overtopping of the Klang river and its tributaries. As already mentioned, the storm of 15 January 1971 caused record floods of 569 m^3 per second; the volume of the river's overflow was 170 m^3 per second in 16 hours and the water receded in 30 hours. The flood damage was around US $34,267,000. The Kuala Lumpur Flood Mitigation Project has undertaken the improvement of the main river canal of the Klang, the Gombak and the Batu rivers in order to alleviate major floods caused by the river. The Project also conducts other activities such as dam construction and upgrading the urban drainage system.

In Malaysia, debris flows down the mountains on the spine of the peninsula, having elevations of over 1,500 metres. Malaysia's mountain/hill areas are wet all year round, recording an average annual rainfall of about 3,100 mm.

Landslides in Malaysia often occur due to the development and building of roads on steep slopes. A particular kind of slide, similar to a mud flow, affects villages and miners. Slope failures cause casualties and injuries. It is reported that the debris flow in Sarawak includes much wood debris from intense logging in the slopes.

VI. KOREA

Korea is relatively free from natural disasters caused by earthquakes and volcanic eruptions but is affected, often severely, by other types of natural disasters such as typhoons, floods, storms, heavy snow, etc. Of these, floods are the most common occurrence. Among the prevailing hazards in Korea are typhoons in the southern part of Korea (Pusan, Masan Yeosoo), heavy rainfall throughout the country, tidal waves on the east coast and heavy snowfall in the middle and western mountain areas.

Two major typhoons, Thelma and Gladys, occurred in southern Korea in 1987 and 1991 respectively. The casualties of the former were greater than those of the latter, while the damage to property stood at around

US $489 million for the former and US $2946 million for the latter. Heavy rainfall was caused by Typhoon Dot, which landed in central China, being weakened and transformed into an extra-tropical cyclone. Observations showed that there was, in the central Korean peninsula, a strong convergence between the south-east wind driven by Typhoon Dot and the north-west wind drawn by the continental high pressure zone.

Korean disasters are attributed to the following:

1. Annual mean precipitation of 1,159 mm is higher than the worldwide mean amount of 750 mm.
2. Rainfall in the summer season (June-September) accounting for as much as two-thirds of the annual amount.
3. Frequent heavy local storms in recent years.
4. Two or three tropical cyclones hitting the Korean peninsula annually.
5. Frequent flash floods due to the osography of the Korean peninsula.

Of all the river basins in Asia, the Han river basin of Korea is of particular interest because of the great flood in the first year of the IDNDR. This flood occurred on 10-11 September 1990, in the middle of the Korean peninsula in which the basin is situated. The rainfall (lasting two days) averaged about 400 mm throughout the basin with a maximum of 584 mm in the upper catchment area, which happens to be a mountainous region as well.

Although this was by no means the maximum rainfall recorded over two days in Korea, it caused the most disastrous flood because the rainfall was confined to a single basin. The metropolitan area and the river-banks, both densely populated areas, were heavily flooded. This was consistent with the heavy rainfall warning issued earlier by the Korean Meteorological Administration (KMA) and the flood warnings by the Han River Flood Forecasting Centre (HRFFC). The flood claimed 157 lives, rendered 187,193 homeless and there were property damages of around US $570 million. The level of the river measured at Seoul was the second highest ever recorded. The first measurements were made in 1900, and the highest river level was recorded in 1925.

The Seoul region, i.e. the northern part of Korea, is badly affected by river floods caused by rain. Obviously there are both structural and non-structural measures for dealing with these water-related natural disasters. The river basins of Han, Makdong, Guem, Seojin and Young San rivers are sensitive as far as processes of forecasting and warning are concerned. Facilities are being upgraded so that the ability to respond to such natural disasters is further improved. The organizations

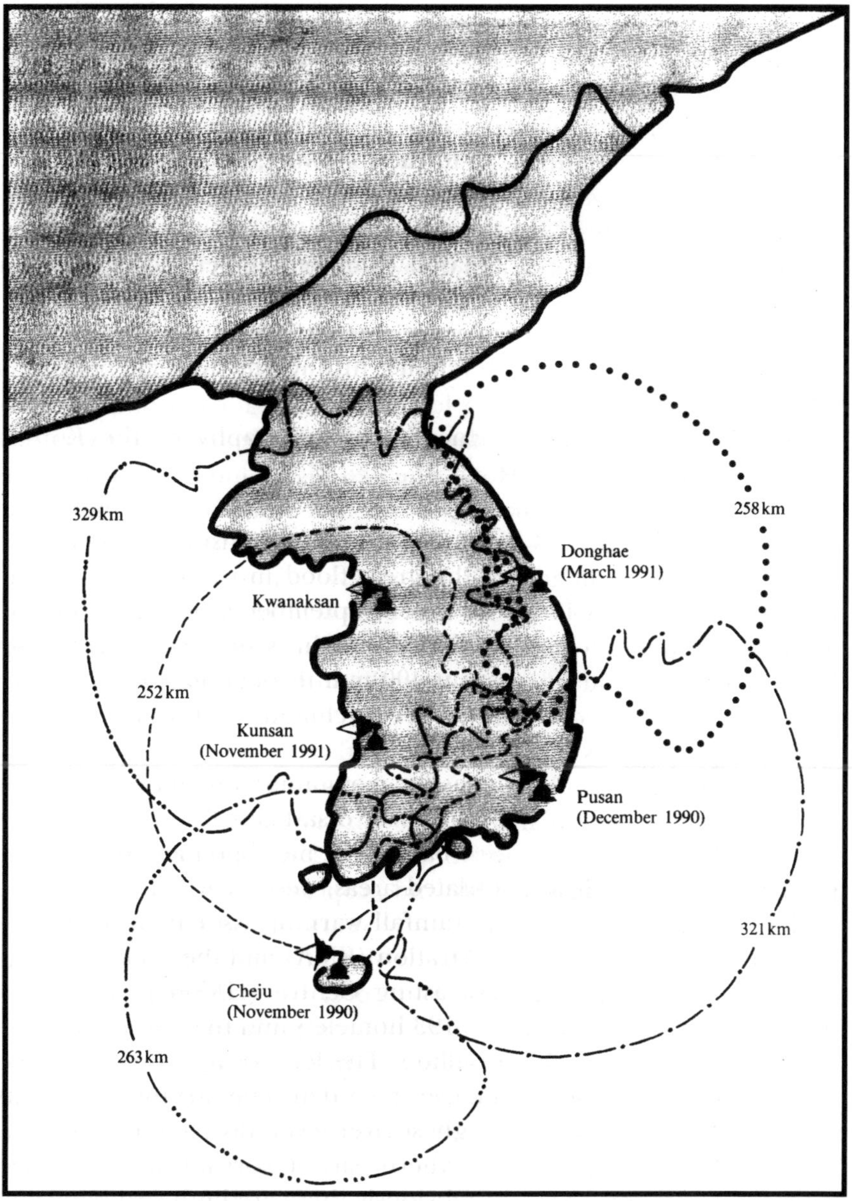

Fig.10 The Korea Meteorological Service radar network: locations of radar sites and their effective coverages. Estimated operation dates of various sites are given in parentheses.

called the Central Anti-Calamity Headquarters, the 15 Provincial Anti-Calamity Headquarters and the County Anti-Calamity Headquarters (established by the ministry) take care of water-related natural disaster prevention, preparedness, countermeasures and rehabilitation programmes. The KMA is responsible for heavy rain warnings. The KMA uses models for Numerical Weather Prediction (NWP) output with regard to rainfall area, rain pattern and rainfall intensity. Although such output has a few limitations, on the whole it certainly helps forecasters to issue early warnings for heavy rainfall. As far as dams are concerned, a regulation stipulates that during the rainy season the water level of each dam should be maintained below a certain level for flood storage.

VII. VIETNAM

Vietnam has suffered appalling losses from all kinds of disasters and around 70 million of its inhabitants have been affected. Two rivers run through Vietnam: the Song Hong (the Red River) and the Mekong. The Song Hong keeps accumulating silt washed away from the northern mountains, and hence its banks have been continually elevated over centuries. During inundations this often poses a threat to the population because the river flows 17 metres above the level of the valley. Typhoons, floods and frequent torrential rains afflict the Vietnamese. Typhoons, floods, droughts, landslides and tornadoes occur at least a year, storm surges occur every two to three years; except in the Hdower Mekong basin, minor droughts occur every two to three years or every five to ten years. While typhoons, floods, storm surges and tornadoes affect both urban and rural residential areas, drought and landslides affect mostly agricultural areas. It goes without saying that water-related natural disasters cause heavy losses of life and property.

The country suffered in the early years of the IDNDR from the October floods and Typhoon Angela that affected 230,000 people in the Mekong delta. A tidal wave swept over two provinces affecting a further 14,000 people who lost most of their property. On 5-9 October 1992, heavy floods caused by massive downpours throughout the central provinces affected 68,360 residences, 13,000 hectares of rice and 12,000 hectares of other crops. 55 persons were killed in the central provinces. Three weeks later heavy floods affected 58,000 persons in the districts of Quang Ninh and Le Thuy. The average rainfall is about 400 mm, but Truong recorded 616 mm. In Son Commune nearly 3,000 people had to be evacuated because of floods. In many areas the water was

2.5-3 metres deep. In 1984 and 1985, the country was repeatedly struck by typhoons and floods.

Wind storms usually accompany typhoons, and heavy rainfall may also occur in June-November. Localized wind storms are also likely to occur during April -May. In the delta, wind storms causing a rise in the water level can cause heavy damages. The coastal areas of northern and central Vietnam are prone to storm surges.

Drought occurs annually in the lower Mekong basin. Even during the rainy season droughts occur due to long dry spells. The severity of droughts is reinforced by the acidity and salinity of the soil. The coastal plain of the Mekong delta suffers terribly from drought during the dry season; there is a shortage of water for human consumption and agricultural needs. Droughts are monitored by the ministry of water resources in the major delta and control region of the country. The intrusion of salt water damages about 1.2 million hectares of the delta.

The Mekong river originates from the Tibetan plateau at an elevation of 5,000 metres. It flows more or less in a southern direction through or along the border of six countries: the People's Democratic Republic (PDR), the kingdom of Thailand, Cambodia and then Vietnam. In fact, it flows through southern China and north-eastern Myanmar before entering the lower Mekong basin. The Mekong is the longest river in South East Asia and of course, one of the largest rivers in the world. It ranks twelfth in terms of drainage area (795,000 sq km) and eighth in terms of run-up (475,000×106 m^3).

The lower Mekong basin covers almost the whole of the Lao PDR and Cambodia, one-third of Thailand and one-fifth of Vietnam, with a drainage area of 606,000 sq km. The lower Mekong basin is located in the world's hazard belt of wind storms, storm surges, floods, droughts and earthquakes but the major disasters occurring there are floods and droughts and to a lesser extent wind storms.

The lower Mekong basin witnesses around three to four storms out of an average of 31 cyclonic storms originating from the western North Pacific and South China Sea. If there are more than six storms, floods will occur in the basin. There are less than three droughts per year. Floods occur when two or more storms occur in succession. Almost immediately after the equatorial trough zone (the foremost part of the south-west monsoon) passes through one of its virulent stages, there invariably ensues a tropical storm.

The Mekong has an enormous volume of excess water during the wet season, hence severe flooding and damages occur almost every year on the fertile floodplain along the main stream, major tributaries as well as the vast floodplain of the delta. The primary flood problems in the lower Mekong occur in three areas: (1) Vientiane-Nong Khai floodplain; (2) the area surrounding Cambodia's Great Lake and (3) the Mekong delta of Vietnam. As is well known, the magnitude and extent of flooding in the flood-prone areas differ because of local conditions and distribution of rainfall. According to an ESCAP publication, in the upper reaches from Chiang Saen (Thailand) to Thakhek (Lao PDR)/Nakhon Phanom (Thailand), every key station recorded the maximum discharge during the 1966 flood; in the middle reach in Savannakhet (Lao PDR)/Mukdahan (Thailand) and the Pakse (Lao PDR), the maximum discharge recorded was during the 1978 flood; at Kratic (Cambodia), the highest recorded flood was in 1939, and at Chau Doc in the Mersong delta (Vietnam), the maximum water level was recorded in 1961.

In the upper reaches some areas are subject to direct flooding from the Mekong river on both banks of the main stream. The Vientiane/ Nong Khai plain in the Lao PDR/Thailand is the worst affected area. Between 1886 and 1966 floods in Malaysia were estimated to have caused damages amounting to US $22 million.

Two-thirds of the city of Vientiane and the whole town of Nong Khai were inundated for almost a month. In the middle reaches the flood-prone areas are along both banks of the river from Thakhek/Nakhon Phanom to Pakse. The worst flood in this area occurred in 1978. At Pakse the water level was 2.5 metres above the flood stage and almost the entire town was flooded.

In the lower reaches in Cambodia, where the land is low and flat, a larger area is usually inundated. Flooding may be attributed not only to the Mekong, but also partly to the backwater effect from the Great Lake. The frequency of flooding is more than that observed upstream. In Vietnam a large part of the Mekong delta is often flooded for extended periods. During the 1966 flood, the damage in the lower reaches of the Mekong delta was estimated at US $34 million.

Over the years, the lower Mekong river basin has assumed enough importance to become a site for international integrated river basin planning. This was the rationale behind the UN's establishment of the Mekong Committee in 1957, consisting of four lower riparian states: Thailand, Cambodia, Laos and Vietnam. The Committee deserves credit for the flood forecasting and warning system since 1970 and one may aver that floods might have caused more deaths and damage without the efforts of the Mekong Committee. The Committee's flood damage mitigation programme worked on small-scale projects and operational hydrology that seldom require large organizational structures or heavy investment.

The severe floods of 1966 called for a flood warning system and in January 1968 the Mekong Committee sought the assistance of the UNDP for establishing a basin-wide flood forecasting system for the lower Mekong basin. A report of the endeavour was published in December 1969 and was approved by the Mekong Committee in 1970. The forecasting programme worked on the feasibility of a centralized forecasting system using advanced computer techniques and mathematical models for the 1970 Mekong flood season. The

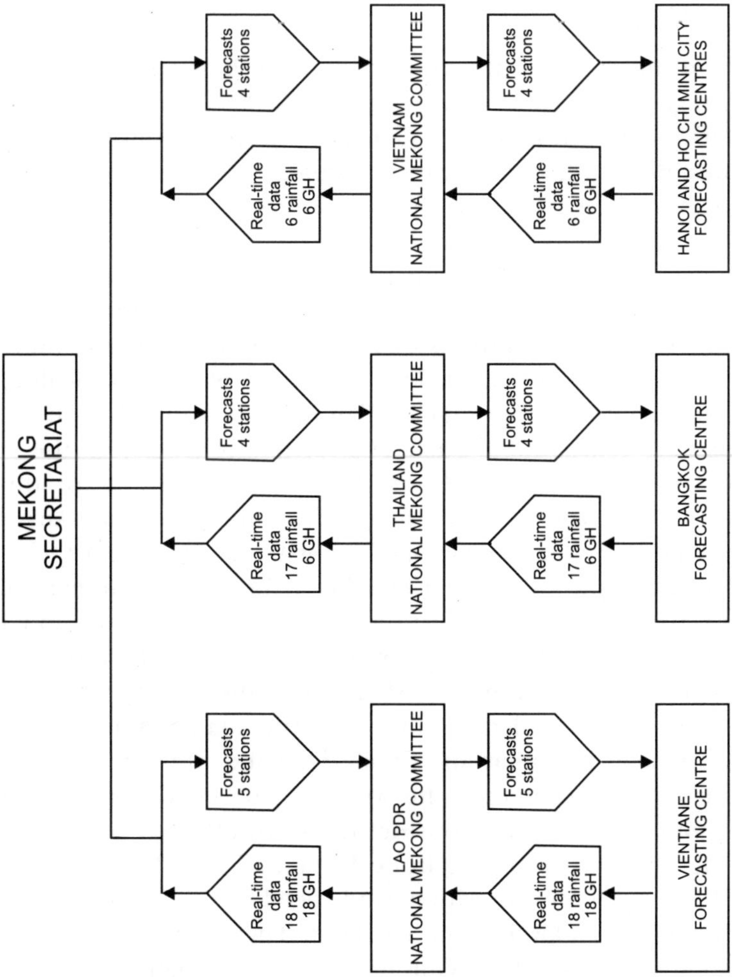

Fig.11 Dissemination of forecasts in the Mekong Basin

simulation of flood forecasts was conducted through three models: (a) SSARR (Streamflow Synthesis and Reservoir Regulation) for the upper and middle reaches; (b) DELTA for the lower reaches of the Mekong delta, without tidal effects; (c) TIDAR for the main channels of Mekong estuaries with tidal effects. These models present flood forecasting operations in their entirety. The report includes details of scientific experiments and the equipment used; it also details the forecast of the flow of a major tributary into an existing reservoir. The flood forecasting operation in the Mekong delta and the dissemination of forecasts in the Mekong basin are shown in Fig. 11.

VIII. THAILAND

Thailand, like most of the countries in the East and the South East Asian region, experiences meteorological and hydrological natural disasters. Natural disasters in Thailand are mostly caused by tropical cyclones, floods, droughts and earthquakes along with of some of its offshoots, such as tsunamis. Other disasters such as landslides, debris flows and mud flows have rarely occurred.

Thailand is struck by about four TCs each year. In reality, such cyclones dissipate into a tropical depression while entering Thailand. TCs provide large quantities of rain becoming naturally useful for agriculture. Any statistical study spanning a few decades, dealing with the frequency of TCs entering Thailand will show that the worst effects of TCs were felt in 1983. Such TCs really originate from the Pacific Ocean and/or the South China Sea and move westward through Vietnam, Laos, Cambodia and finally Thailand. From June-August the mean track of cyclones lies across the north-eastern and northern parts of Thailand; from September-October it lies across the central and eastern parts, the upper portion of southern Thailand and lower portion of the north-east; in November-December it travels further south from Chunphon. In April-May, most of the cyclones originate over the Indian Ocean and move over the Andaman Sea and the Bay of Bengal towards Myanmar and finally Thailand. During these months, the cyclones affect mainly the western part of Thailand. Bangkok Metropolis and its neighbouring areas are affected by the cyclone entering Myanmar, resulting in heavy rainfall and flooding.

In 1989 Typhoon Gay, with a wind speed of 120 kmph at its centre, was the first of its kind to hit Thailand. The 1989 cyclone originated in the lower portion of the Gulf of Thailand, 270 km east of Songkhla on 1 November. At 03:00 hours on 2 November, the depression was transformed into a tropical storm, and the next day at 16:00 hours, it assumed the form of a typical typhoon. Typhoon Gay moved west-north-westward towards the upper portion of southern Thailand and the border of Churthon and Prachuat Kirikhan, and while landing struck two districts of Chunthun at around 22:30 hours on 4 November. It caused severe damage to property and claimed lives in the southern and some eastern parts of Thailand.

Drought, another type of atmospheric disaster, occurs around May and sometimes before the beginning of the rainy season. Upper Thailand, especially the northern and north-eastern part of the country, is susceptible to drought.

Thailand often experiences earthquakes with epicentres within or outside the country. Thailand recorded an earthquake measuring 5.9 on the Richter scale on 22 April 1983 at the Srinagar end dam, Kanchanburi province, about 200 km from Bangkok. On 6 November 1988, another earthquake occurred several thousand km from Bangkok in the south of China, measuring 7.3 on the Richter scale. The impact of the earthquake could be felt, not merely outside the country on the Thai-Myanmar Border, but even as far as Lao PDR and the Andaman Sea. The Thai Meteorological Department has a station-based network including a Seismic Research Observatory (SRO), Worldwide Standard Seismograph (WWSS) network and Short Period Vertical Seismograph (SPVS) station. Of course, there are also government agencies for the preparedness and mitigation of earthquake disasters.

Bangkok, the capital of Thailand, located in the southern part of Chaw Phraya river delta, is a low-lying flat plain. The general impression is that while in the past the inhabitants of this region were used to floods and could put up with them, the subsequent spate of urbanization has adversely affected drainage efficiency and flood storage. Arguably, one can identify a variety of factors such as (a) rainfall, inclusive of direct precipitation, inflow from the outer area and river flow, (b) tides, (c) land use and (d) town planning. At present steps are being taken to protect Bangkok against floods, with structural and non-structural measures including land-oriented systems, diversion- or water-oriented systems. Flood forecasting and warning systems cover Pasak Nan, Prachin, the Buri river basin and the Bangkok Metropolitan Area. Computer and satellite image receiving sets are extensively used. The Meteorological Department and the Royal Irrigation Department are responsible for flood warnings in the country and the Department of

Drainage and Sewerage of the Bangkok Metropolitan Administration takes care of Bangkok City. The first two departments are also responsible for drought forecast and warning.

Fig.12 Flood in the eastern part of Bangkok Metropolis (by courtesy of Meterological Department of Thailand)

There is a variety of methods for storm surge forecasting and warning, but for typhoon warnings there is a sharing of facilities such as telecommunication links with the WMO and with other countries, satellite image receiving stations, radars and computers.

On 26 December 2004, the Sumatra-Indonesia earthquake hit the shores of Thailand, killing a large number of foreigners. Keen to develop future facilities, Thailand quickly responded to the situation: on 28-29 January 2005 a meeting was convened at Phuket for developing an early warning system for tsunamis in the Indian Ocean. The deliberations at Phuket led to the declaration on international cooperation for setting up an early warning system for tsunamis in the Indian Ocean. Effective steps following this declaration were to be taken by mid 2006.

IX. SRI LANKA

Sri Lanka is on the main sea routes of the Indian Ocean stretching from the Middle East through the Malay Straits to China. An ancient South

Asian civilaztion, Sri Lanka's past yields cultural and historical evidence of major innovations in trade and technology. In his cardinal treatise *Science and Civilization*, Prof. Joseph Needham described the fusion of several civilizational patterns that 'took place in Ceylon, the work of both Singhalese and Tamil'. The construction of reservoirs, irrigation canals and large edifices in Sri Lanka is comparable to the architectural feats of ancient Egypt, but many of these structures in Sri Lanka were destroyed by natural disasters. The worst destruction occurred in the thirteenth century when a tectonic movement caused the Mahaweli River to change its course drastically. Sri Lanka is mostly subjected to natural hazards such as cyclones, floods, droughts, landslides and coastal erosion. Its coastal lines make it especially vulnerable.

Sri Lanka is located on the western side of the entrance to the Bay of Bengal and it is thus well within the domain of cyclones that develop in the bay. Also, Sri Lanka's location between 5°55' and 9°55' North and 79°40' and 81°53' East places it within the monsoon wind belt. Therefore, there are ample meteorological reasons for heavy rainfall during the south-west monsoon period (October-February) and south-east monsoon period (May-September). Floods occur in the south-western plains almost every year affecting hundreds of people. The highest point of the mountain ranges (in the south central area) is 2,526 metres above the mean sea level. The elevation decreases towards the plains of the west and south-west and also towards the north.

The eastern and north-eastern coasts are affected by rather unusual cyclones. The last (and perhaps the most destructive) one occurred in 1978 claiming 940 lives and rendering innumerable people homeless. During end May and early June 1989, torrential rain caused a devastating flood coupled with landslides. The latter occurred in the upper reaches of the three major river basins, Kelani Ganga, Kalu Ganga and Gin Ganga in the south-western part of Sri Lanka. These natural disasters wreaked havoc in ten districts killing hundreds of people, displacing thousands and damaging property worth US $30 million. Although floods and landslides due to monsoon rains occur more or less annually, the worst incident was certainly the one in May-June 1989.

The agencies that deal with forecasts and warnings for bad weather and floods are the Department of Meteorology and the Department of Irrigation respectively. In order to protect Colombo (the capital of Sri Lanka) from the annual flood water of Kelani Ganga, the government has a scheme to (a) minimize flood damages by prior warning (b) to safeguard the flood protection work and lessen the suffering caused by floods. Flood levels are identified. They are the

(a) minor flood level, i.e. the water level in the river is five feet (1.75 metres) (b) major flood level, i.e. the water level rises to eight feet (2.75 metres) (c) dangerous flood level, i.e. the water level reaches ten feet (3.3 metres) (d) critical flood level, i.e. the water level reaches 12 feet (4 metres). Such flood levels are read on the Gange Station on the banks of the Kelani Ganga. The flood protection work in the Kelani Ganga basin was begun in the eighteenth century by the Dutch. Later, during the twentieth century, the British built the North and the South Bunds as they stand today. No development has taken place thereafter. The Kelani river basin has a network of hydrometric stations. There exists an efficient mechanism for communication and warnings.

The Government of Sri Lanka has undertaken several programmes to generate information on the causes of disasters and the limitations of the warning system, communication system, etc. The guides and manuals published by the government are very useful.

As mentioned earlier, the rainfall in Sri Lanka is determined by the monsoon winds which originate either over the ocean or over the Himalayan range. There was thought to be a relationship between rainfall intensity, landslides (including mudslides), floods and seismic activity in the upper catchments of the Mahaweli River for a given period. However, recent studies show that most of these floods in the Mahaweli catchment area are absorbed in the reservoirs of the Mahaweli and it is now clear that there is no direct relationship between landslides and seismic activities. Places where landslides occur are identified as high-risk areas, and measures are being taken to prevent landslide disasters in these areas. There are some identifiable areas for disaster mitigation in regions prone to landslides and preventive measures are being taken through afforestation (with fast-growing trees that have high evapo-transpiration) in order to reduce the soil moisture in these areas and thereby prevent landslides. The vulnerability of areas subject to inundation needs to be identified clearly before evacuation can be carried out. The Mahaweli project, therefore, has been given high priority by the government and by NGOs. Sri Lanka has always sought the cooperation, collaboration and support of other countries having the necessary expertise and infrastructure.

The most recent disaster causing widespread death and destitution in Sri Lanka was the tsunami on 26 December 2004. As the region is wholly surrounded by coastal lines, hardly anything was spared by the tsunami (which had travelled from the Indonesian archipelago). The event reiterated the extreme vulnerability of the high-risk zone to water-related disasters.

X. PAKISTAN

Pakistan lies between 23° and 37° North latitude and 61° and 67° East longitude, and has a total area of 888,000 sq km. The temperature variation in the country ranges from 6°C to 49°C. The monsoon climate is dominant, thus the average rainfall during the monsoon season is around 200 mm. Pakistan has the second highest peak in the world, with major mountain ranges—the Karakoram and the Hindu Kush—in the north. There is also a vast arid desert in the south. Thus Pakistan has areas prone to disasters such as floods, drought, earthquakes and landslides. As floods constitute the principal natural disasters, they have engaged the attention of researchers of river basin management.

The Indus is the country's main river system. The Indus river basin consists of five major rivers that flow in a southern direction to the Himalayan ranges and converge to form one river, approximately halfway between the mountains and the Arabian sea. The five tributaries of the Indus—Jhelum, Chenab, Ravi, Sutlej and Beas—are situated roughly on the east of the Indus' main stream. The Indus river and its tributaries rise in the glaciated mountain of Southern and Central Asia and flow through parts of Tibet, India, Pakistan and Afghanistan. About 60 per cent of its total drainage area lies in Pakistan. The continental climates of Central and Western Asia influence the mountainous upper basin. The climate is characterized by a westerly pattern of circulation, late winter snowfall, cold winters and hot summers. The headwater of the Indus is located high in the mountainous region. West of the Indus, the Kabul river and its major tributaries such as the Swat river also originate in glaciated areas. The headwaters of the Indus' five tributaries are fed by glaciers and melting snow but they are also influenced by the summer monsoon circulation and associated rainfall. The influence of the monsoon is felt as one moves from west to east; e.g. the Jhelum receives its monsoon rain only after it has left the Kashmir valley and its waters have been stored in the Mangla dam, Pakistan's largest reservoir. The other four tributaries feel the influence of the monsoon after leaving the Siwalik foothills, flanking the Himalayan ranges. While the upper reaches of the Indus and its tributaries descend into the plain, the climate, landscapes, peoples and the water system of the Indus plains differ remarkably from those of the upper basin. The plains have hot summers, mild winters, short monsoon winds and precipitation around the months of July and August. The north-eastern part of the Indus is situated in Punjab and the North West Frontier Province (NWFP) in the west; thus Lahore experiences the highest average annual rainfall, there is 180 mm of rain in Multan (to the south) and 330 mm in NWFP. The famous and historic city of Taxila, just north of the Potowar plateau of central Punjab is irrigated by a combination of perennial streams and rainfall. Flood farming or recession agriculture (locally known as *sailaba*) has been conducted since antiquity and is still carried out on some 1.23 million hectares. The lands between the floodplains have been well irrigated since Mughal times. The irrigation system was extended and improved by local rulers, princes and the British.

The mountains of Baluchistan help develop micro-climatic and micro-hydrologic niches for hilly terrain. The small-scale canal floods in recent times are perhaps more frequent because of reduced flow passage caused by irrigation and power projects, causing alluvial deposition and at some places elevation of the riverbed above the adjacent floodplain. The floods of 1988 have had few parallels in the history of Pakistan. They occurred in two phases over a period of 80 days. With the protective measures, flood forecasting and warning systems, the first phase of the flood was expected. The unexpected phase of the flood occurred in the southern part of Pakistan, where a 100-metre wide breach was created in the protective embankments on the left bank of the Indus river on 17 August 1988. The breach widened to 600 metres and approximately 162,000 hectares of land were inundated. About 100,000 people needed rehabilitation. A much more dangerous situation arose on 25 September 1988 because of continuous heavy rainfall in the catchment areas. This second phase was all the more distressing because of the return of the evacuees of the first stage, who had barely recovered from their earlier shock. This was attributed mainly to an information gap and the lack of early warning to the public, perhaps because of dysfunctional equipment (such as radar, telemetric gauges, etc.). The floods of September 1992 were caused by heavy rains in the upper reaches of the Jhelum river. The discharge of excess water into the main stream of the Indus debilitated the major barrages of that river. Even the Mangla dam could hardly withstand the monsoon run-up. Total damages were around US $1 billion, including US $513 million in public infrastructure damages. The violent swell of the Jhelum and Chenab rivers seriously affected the NWFP and the southern province of Sind.

Though earthquakes seldom occur in Pakistan, an earthquake measuring 7.5 on the Richter scale took place in 1935 in Quetta, killing around 35,000 people. The area is still considered to be an active seismic zone. Since its creation Pakistan has encountered five major earthquakes. One of them, measuring 6 on the Richter scale, struck the northern areas of Pakistan in the Karakoram range on 28 December 1974. There were about 1,000 casualties, nearly 2,000 people were disabled, 18,600 houses were destroyed and 2,500 houses partly damaged. About 65,000 livestock were lost. Rockfalls, landslides and even subsidence occurred on the Karakoram highway adjacent to the Indus. An important consequence of such natural hazards is the establishment of a chain of seismological stations with its headquarters at Quetta, run by the Pakistan Meteorological Department, which often seeks the collaboration of WMO to strengthen its network and a seismic array system at Islamabad.

There are several agencies besides the Pakistan Meteorological Department for organizing disaster management from the federal to the provincial/district levels. The Federal Flood Commission and the Forecasting and Warning Centre mitigate the damaging effects of floods and implement relief programmes. An Emergency Relief Cell has been set up at the federal level with counterparts at the district levels. The Army has a far greater role to play than merely providing aid to civil administration; e.g. during the annual flood season the Engineering Directorate of the Pakistan Army operates the flood relief control centre at the General Headquarters for monitoring the flood situation and coordinating relief and rescue operations involving Army units. The aforementioned Flood Forecasting and Warning Centre works with these units at different levels. As elsewhere, the Army undertakes emergency repairs of the communication and transport network during floods. On many occasions Pakistan has also sought foreign expertise and financial support.

XI. BANGLADESH

Bangladesh is located between South and South East Asia. It extends between 21° and 27° North latitude and 88° and 92.5° East longitude. The Bay of Bengal is to the south of the country. It is essentially a deltaic alluvial plain formed by the Ganges-Brahmaputra-Meghna river system. The first two rivers originate in the Himalayan ranges and after travelling through India and Tibet they flow into the Bay of Bengal. Bangladesh is prone to natural disasters—because of its geographical location in the cyclone-prone area it experiences frequent TCs arising from tidal surges. These have caused enormous loss of life and damage to property. Besides TCs, Bangladesh is also subject to tornadoes and (both long-lasting and flash) floods. Droughts occur at intervals of a few years, sometimes a decade.

Cyclones originate in the Bay of Bengal. Cyclonic storms are caused when winds and atmospheric pressure drop to a subnormal level. They often hit the southern areas of the country between April and December. The cyclone of 1988 with tidal surges along the Bay of Bengal killed 5,000 people and 0.1 million animals and caused enormous damages to agricultural crops and forests. The tidal surges accompanying the cyclones possess a lethal velocity, often reaching a height of around 20 metres. The sea inundates not only the coastal areas but also the inland areas; inundation by sea water makes the soil saline. The South Asian Disaster Management Centre (SADMC) analysed the devastating cyclone of 1991 thoroughly, as a result of which cyclone warning and mitigation measures were undertaken in earnest. Such studies reveal the necessary conditions for cyclones. The cyclone that struck Bangladesh on 29-30 April 1991 was particularly severe, causing around 138,882 fatalities. The total loss to lifeline systems and property was estimated at US $2.07 billion. In Bangladesh a complex delta is formed by the sedimentation of silt from the Ganges, Brahmaputra and Meghna. The coastal morphology of Bangladesh is characterized not merely by a network of rivers and silt deposits, but also by a fairly large number of islands. Furthermore, a submarine canyon runs south-east/north-east, partly across the continental shelf, about 24 km south of the Bangladesh coast. The coastal areas of the country are situated north of the funnel-shaped and shallow Bay of Bengal. The height of tidal surges around the coast of Bangladesh ranges from 5-6 metres. The whole coast is about 710 km long, having distinctive geo-morphological features in the eastern, central and western regions.

The atmospheric environment is also distinctive. One hardly finds strong vertical or horizontal winds near the centre of the cyclone, but a strong vertical shear of opposite signs is present on either side of the system. The difference between the wind vectors between two vertical levels is called the vertical wind shear. This limits the number of cyclones to about one or two per year in April-May and mid October-mid December. In terms of fluid dynamics, there is a low pressure region with cyclonic vorticity and warm ocean temperature. There is scarcely any tropical cyclone in the sea if the temperature is less than 27°C. The high surface temperature is necessary for a sleep lapse rate for maintaining the vertical circulation in a cyclone. This is by and large a prerequisite for cyclones in the Bay of Bengal. The cyclonic storm of April 1991 started off with a core of hurricane winds first detected on April 23 in a low pressure area near the Andaman Sea. As the storm remained stationary for two days it acquired energy and was concentrated into a depression in the early hours of 25 April; it then began to move somewhat slowly in a west-north-westerly direction, but developed into a deep depression on the same day. At midnight 26 April the depression concentrated into a cyclonic storm once more and moved northwards. It intensified into a severe cyclonic storm at midnight 27 April; it continued to intensify till the morning of 29 April, after having moved close to Chittagong at midnight. Finally the cyclone crossed the Chittagong coast, a little north of Chittagong at 2 am on 30 April 1991. The cyclone being thus guided by the steering wind with its anti-cyclonic flow pattern at 200 hpa, its future path and landfall area could be estimated. One could also compute the central pressure of the cyclone (940 hpa) and the pressure drop (69.5 hpa); the wind speeds could be calculated and they agreed with the forecast speed. Since 1970, the meteorological office of Bangladesh has had efficient systems to predict the intensity and course of a cyclone; its space research centre collects satellite imagery of the cyclone and its movement and electronic media are used to disseminate warnings well in advance. This is necessary as several millions of people live in the cyclone belt and their houses can hardly withstand cyclones of high intensity. Since 1972 the Government of Bangladesh has implemented the Cyclone Preparedness Programme in collaboration with various international bodies and organizations.

The April 1991 cyclone caused extensive damage to the agricultural sector, especially to standing crops, livestock, forestry and fishery resources. On the occasion of the April cyclone, the meteorological office in Dhaka transmitted the message about the development of the cyclone, and the volunteers were informed on 27 April 1991 when the storm concentrated into a severe cyclonic storm with a core of hurricane intensity. Though a lot of public warnings were given, it was only at 1300 hours on 29 April that evacuation actually began. The storm struck Chittagong and the Cox Bazar coast at 1800 hours with wind speed rising to 225 kmph at midnight. The storm passed the coast in the early hours of 30 April 1991. Around 50,014 people were killed, to say nothing of the loss of cattle and livestock. There were extensive relief and rehabilitation efforts and several Cyclone Centres were established. The process of tracking and forecasting the April cyclone and the relief measures that followed provided Bangladeshi organizations valuable experience for mitigating disasters, minimizing losses and restoring normalcy to affected regions more efficiently in the future. A critical review

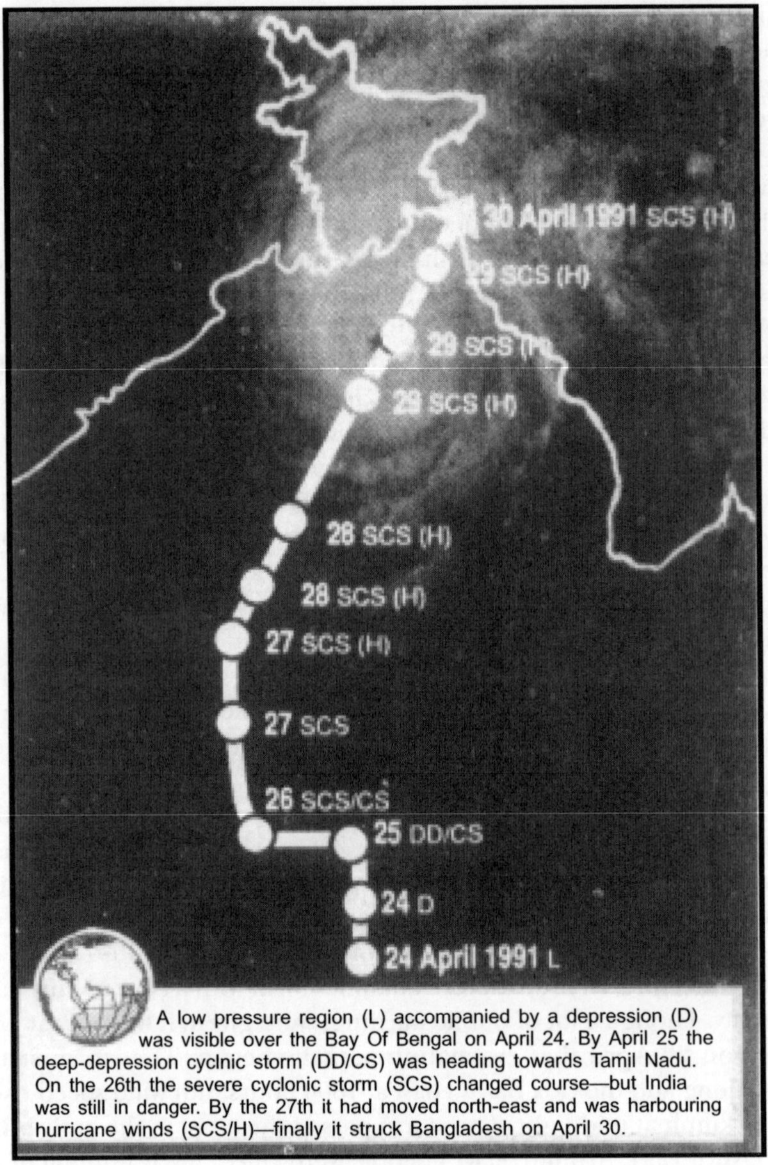

A low pressure region (L) accompanied by a depression (D) was visible over the Bay Of Bengal on April 24. By April 25 the deep-depression cyclnic storm (DD/CS) was heading towards Tamil Nadu. On the 26th the severe cyclonic storm (SCS) changed course—but India was still in danger. By the 27th it had moved north-east and was harbouring hurricane winds (SCS/H)—finally it struck Bangladesh on April 30.

Fig.13 Fighting the elements

undertaken by them in the aftermath of the April cyclone has done much to develop ways of protecting inhabitants of the coastal belt from the effects of cyclonic disasters.

Floods are almost as devastating as cyclones in terms of their magnitude and intensity. Because of its predominantly riverine terrain,

Bangladesh appears to be capable of coping with ordinary floods by controlling them locally and thus protecting communities. The rivers of Bangladesh drain the run-off from a catchment area of about 1.7 million sq km; about 7.5 per cent of this run-off is drained from the country.

In years when rainfall in the catchment areas of the river system is not extraordinary or when the peak floods of the river systems do not synchronize, nearly 20 per cent of the country is flooded. When there are severe floods, 55-60 per cent of the country is submerged.

The following are held to be the major causes of floods in Bangladesh:

1. Extraordinary rainfall in the catchment areas of riverine systems.
2. Flat topography and low elevation of the country.
3. Spill from the main river system and from smaller rivers including tributaries of the main river system.
4. Drainage congestion, floodplains and tidal effects.
5. Coincidence of peaks of floods of major rivers.

Although scarcely a year goes by without floods, their magnitude varies from year to year. According to reports, severe floods affect one-third to two-thirds of the country, damaging crops, roads and bridges and often washing away other kinds of infrastructure. In the aftermath of floods, people find themselves without access to food, shelter, etc.

The Government of Bangladesh plays a key role in disaster mitigation and especially flood disaster management. It has issued 'Standing Orders for Floods' to be implemented by the relevant ministries. In order to defend several areas against floods, millions of taka have been invested to construct long stretches of embankments, thereby improving the capacity of thousands of km of drainage channels. A number of agencies actively participate in operations of relief and rehabilitation, often in collaboration with international organizations such as the UNDP office in Dhaka. NGOs are also mobilized to facilitate the participation of the community during and after natural disasters.

XII. MYANMAR/LAO PDR/CAMBODIA

In South East Asia, some common factors pertain to the management of natural disasters. Myanmar, Lao PDR and Cambodia experience frequent floods, drought, cyclones and storm surges. Both structural and non-structural measures have been taken against floods. Meteorological and hydrological agencies provide forecasts and warnings besides supervising rescue and relief work.

Myanmar has a different way of battling disasters because of the nature of the country's administration. The principal hazards that Myanmar experiences are storms, TCs, storm surges mainly on the Myanmar coast (90 per cent on the Rakhine coast, 7 per cent in the delta, 3 per cent on the Tamiathary coast), strong winds over the inland rivers and streams, floods in the catchments of major rivers (Ayeyarwadi, Thanlwin, Bago and Sittaung) and tributaries (Chinbwin and Myitnge), landslides and earthquakes, particularly tectonic earthquakes in the Alpine-Himalaya Earthquake Belt. The Lao PDR is affected by typhoons and landslides.

The floods in Lao PDR are caused chiefly by the overflowing of the Mekong in its upper and middle reaches during the south-west monsoon season (August-September) while the floods in the tributaries are caused mainly by tropical depression or tropical storms. Flash floods occur in many small steep catchments in the northern and southern regions. The frequent flooding of the Vientiane plain (including the city of Vientiane) by the Mekong and its tributaries often causes damages worth tens of millions of dollars. The municipality of Vientiane is responsible for flood protection, while the Central Government, particularly the Ministry of Agriculture and Forestry, is the main agency responsible for flood loss prevention and management.

XIII. AFGHANISTAN

Afghanistan is surrounded by high mountains and is often described as the 'sky's theatre' formed by the high peaks of the Hindu Kush, with altitudes of about 7,700 metres. The country is affected principally by floods, debris or mud flows that occur at least once every year. The northern part of Afghanistan experienced an earthquake on 1 February 1991, causing the deaths of more than 100 people and rendering around 2,000 families homeless. Severe floods during the next two days killed around 400 people and rendered 27,000 homeless. The Hindu Kush were hit on 3 September 1992 by cloudbursts which resulted in mud floods. These caused rivers to overflow, devastating the Salang, Ghordoand and Shutul valleys. The town of Gulbahar, 70 km north of Kabul, was badly affected. 450 people were killed and more than 500 went missing (reportedly those who had returned to Afghanistan from Pakistan as refugees). Urban and rural residential areas are affected by floods, debris and mud flows and floods mostly occur in the north, west and sometimes in the south. There are two kinds of droughts: the first occurs every 5-10 years, while the second occurs every 10-20 years.

Although Afghanistan does not seem to have the necessary infrastructure for disaster prevention, preparedness and mitigation, international support and cooperation have been extended when required. Afghanistan has independently set up the Office of Disaster Preparedness since the beginning of the IDNDR.

XIV. NEPAL

Being surrounded by mountains, Nepal is affected by floods, debris, mudflows and landslides at least once a year but other disasters like earthquakes, droughts, lightning, hailstorms, glacier motion and hurricanes occur less frequently. Floods occur in the Koshi and Bagmati river basin and the Gandaki and Kamali river basins.

Nepal's geographical position on the southern slope of the Himalayas and its fragile soil make it susceptible to disasters. Nepal is included within the range of seismic waves as a result of which the country experiences earthquakes. Earthquakes of moderate and high intensity have occurred at least twice a year during the last two decades. More than 80 per cent of Nepal's land mass is formed by hills and mountains spanning 800 km from east to west and 110 km from north to south. Three major river systems with their numerous tributaries run from north to south along the steep slopes. The elevation of these slopes ranges from 8,848 metres (the highest point) to 70 metres (the lowest point). The extent of loss caused by earthquakes is greater than that caused by floods and landslides, but the latter occur more frequently. The local communities have not fully attended to the consequences of natural disasters. Of course, there are agencies (mostly governmental) that coordinate the activities of the Departments of Hydrology and Meteorology. It is part of Nepal's official policy to seek help from international organizations and donor countries.

XV. INDIA

Natural disasters in India, particularly the ones with a slow onset, occur regularly. The recent tsunami raised the issue of coastal vulnerability, and heavy downpours in Mumbai in 2005 have exposed urban vulnerability. In addition to cyclones in Orissa and seismic activity in the Latur and Bhuj areas, regular floods occur in different parts of the country. The violent earthquakes in Assam and north Bihar will not be easily forgotten.

TCs originate in the Bay of Bengal, the Arabian Sea and the Indian Ocean. Geographically, the region is a slight distance away from the equator and lies within the tropical belt between 30° North and 30° South. The ocean water is warm. The frequency of TCs, their intensity and impact vary from place to place; even though the intensity or the frequency may not be particularly high in the initial stages, it usually increases manifold when the TC crosses the coast bordering the northern part of the Bay of Bengal (i.e. the coastal areas of north Orissa and West Bengal). Low pressure areas in the atmosphere allow the formation of TCs. They extend horizontally for 500-1,000 km along the surface and 12-14 km vertically from the surface. The India Meteorological Department (IMD) has the dataset on the classification and intensity of cyclonic disturbances. The 'eye' of the cyclone situated in the central region of the cyclone has an average radius of about 20-30 km, but it could rise to 40-50 km when it attains maturity. The eye is generally encircled by a ring of very strong winds, averaging up to 30-50 km off the centre. Winds spiralling anti-clockwise in this region in the northern hemisphere often extend outwards for long distances with speeds gradually decreasing further away from the centre. Glimpses of satellite pictures show these winds to be slow, rapid or often gradual. Vertical winds have an anti-clockwise rotatory motion of about 7-8 km followed by clockwise winds of up to 12-14 km in the northern hemisphere. The eye is usually almost cloud-free but has around it a hollow cylinder of clouds. This zone of clouds is the most dangerous part of the cyclone because the winds therein are extremely intense and bear enough moisture for heavy rain. Clouds outside this zone (as evident from satellite pictures) appear in bands that eventually merge with the clouds of the dangerous zone; one can examine the special nature of these clouds via television as they can easily be detected by satellite and radar photographs.

On an average there are five to six cyclones in the Bay of Bengal and the Arabian Sea, of which two to three are likely to be very severe. There are more cyclones in the Bay of Bengal than in the Arabian Sea. Almost every year, May-June and October-November are the stormiest months. The IMD has records of the frequency of cyclones since 1891. Cyclones have resulted in gales, strong winds, heavy downpours and high storm surges. Being the sudden elevation in the sea level due to a TC, a storm surge is greatly amplified near the coastlines where the water is shallow and the topography of the coast resembles a funnel. The storm surges in the Bay of Bengal have engaged the attention of researchers in India and abroad and valuable case studies have been conducted. Area

Cyclone Warning Centres are located in Kolkata, Chennai and Mumbai and there are Cyclone Warning Centres at Bhubaneswar, Visakhapatnam and Ahmedabad. The IMD office in New Delhi provides assistance to Bangladesh, Myanmar, Maldives, Pakistan, Thailand and Mauritius.[5]

NOTES

1. Leet, L Don, *Causes of Catastrophe: Earthquakes, Volcanoes, Tidal Waves and Hurricanes*, McGraw Hill, London and New York, 1948.
2. 'Indonesia quake', *The Sunday Statesman*, Kolkata, 3 December 2006, p.10.
3. UNCRD in collaboration with PHIVOLCS, DOST, 'Tsunami', *The Challenge of Mindoro: Report of the Quick Response Investigation of the Mindoro Island Earthquake in the Philippines on 15 November 1994*, UNCRD Report Series, No. 11, 1995, p.18.
4. 'Philippine typhoon claims 300, 300 more missing', *The Sunday Statesman*, Kolkata, 3 December 2006, p.10.
5. For a detailed study of natural disasters in India the reader may refer to Sinha, D K, *Towards Basics of Natural Disaster Reduction*, Research Book Centre, New Delhi, 2006, pp.21-75. This section examines floods, droughts, storm surges, cyclones, hurricanes, tornadoes, earthquakes, tsunamis and avalanches. The 'Select Bibliography' of the present work mentions other relevant books by the author, Harsh Gupta and Vinod Sharma. The subsequent chapters of the present work refer to disasters in the Indian context, where necessary.

Chapter 2

The Context of the December 2004 Tsunami

INTRODUCTION

Accounts of the tsunami occurring in the early hours of Sunday, 26 December 2004 still shock listeners/readers. Initially, only a handful of people near the Pacific Ocean were aware of the event. The next day, media activity erupted. Victims of the tsunami encountered difficulties unlike those caused by other disasters. Killer waves are common enough for the Japanese to whom we owe the word 'tsunami', literally a 'harbour wave'. Despite the virtually negligible elevation of the sea level, the fury of the wave that rushed towards the shore took people by surprise.

Broadly speaking, the response to the tsunami had a socio-political bias rather than a regard for the purely human-ecological aspect. Skepticism and positive criticism were articulated, past examples were cited and outrage expressed that the necessary measures for preventing such disasters had not been taken. The media emphasized the resilience shown by the communities worst affected by the tsunami. The Tampere Declaration on Disaster Communications (1991) had earlier drawn attention to the importance of communications in disaster management. An outcome of the Tampere Conference on Disaster Communications, it recognized 'the urgent communications need generated by emergency disaster relief and the longer-term needs of disaster mitigation'.[1] Communications have dramatically changed with rapid advances in technology and mass communication must necessarily draw upon all available forms of assistance. This chapter tries to develop a contextual model of the tsunami, drawing essentially on the media coverage of the event, particularly the reports in the print media.

CHARACTERISTICS OF THE TSUNAMI: THE MEDIA VERSION

The Hindu, based in Chennai, had an obvious edge over other newpapers reporting the tsunami because of its proximity to the Indian Ocean. In the four days following the tsunami, there was a copious amount of information on fatalities, injuries, etc. In its 27 December issue, *The Hindu* described the origin, location and the nature of the earthquake (measuring 8.9 on the Richter scale) off Sumatra in Indonesia and its epicentre, 257 km south-south-west of Banda Aceh, Sumatra. According to the newspaper, this was among the most powerful quakes in the last 40 years, reportedly killing 2,200 people in Indonesia, rendering innumerable people homeless and killing 10,000 in India, Sri Lanka, Malaysia and Thailand. The waves were generated by an undersea quake and the tsunami rose as high as six metres throughout the Indian Ocean and the Andaman Sea. 3,000 people were reported dead in Tamil Nadu, Pondicherry, Andhra Pradesh and Kerala. There were 1,705 deaths in Tamil Nadu alone, Nagapattinam had a death toll of 780, Kanyakumari of 392 and Cuddalore of 290. Pondicherry reported 280 deaths, 211 of which were in Karaikal alone. In Kerala at least 100 people died. The toll in Andhra Pradesh was 84. 131 people were killed on the beach at Chennai. In the coastal district the toll was as follows: Kanchipuram 61, Villupuram 23, Tiruvallur 13, Tirunelveli four, Tuticurin two and Thanjavur one. The coastal region of Chennai suffered heavily. There were strong winds at the San Thome Basilica where Christmas was being celebrated: on the Marina one could see the broken frames of fishing boats and the broken walls of the lighthouse and forensic sciences lab. The police headquarters, the All India Radio building and the Foreshore Estate apartments were in a state of ruin. In the fishing hamlets of Aydhya Kuppama and Patanathakkam, many residents died instantly when the seawater struck the coast. The sea waves were powerful enough to overturn vehicles and boats close to the lighthouse. Reporters noted that tsunami waves hit the area at intervals of three hours. Although thousands tried to flee in any way possible, some organized steps were taken for evacuation. Nevertheless, the rescue operation was disrupted because of the continuous surge of seawater. The search operation for those reported missing was terribly difficult because many people were believed to have been swept away by the waves. The beaches of Kottivakkam, Besantnagar and Tiruvaniniur were lashed by ferocious tidal waves.

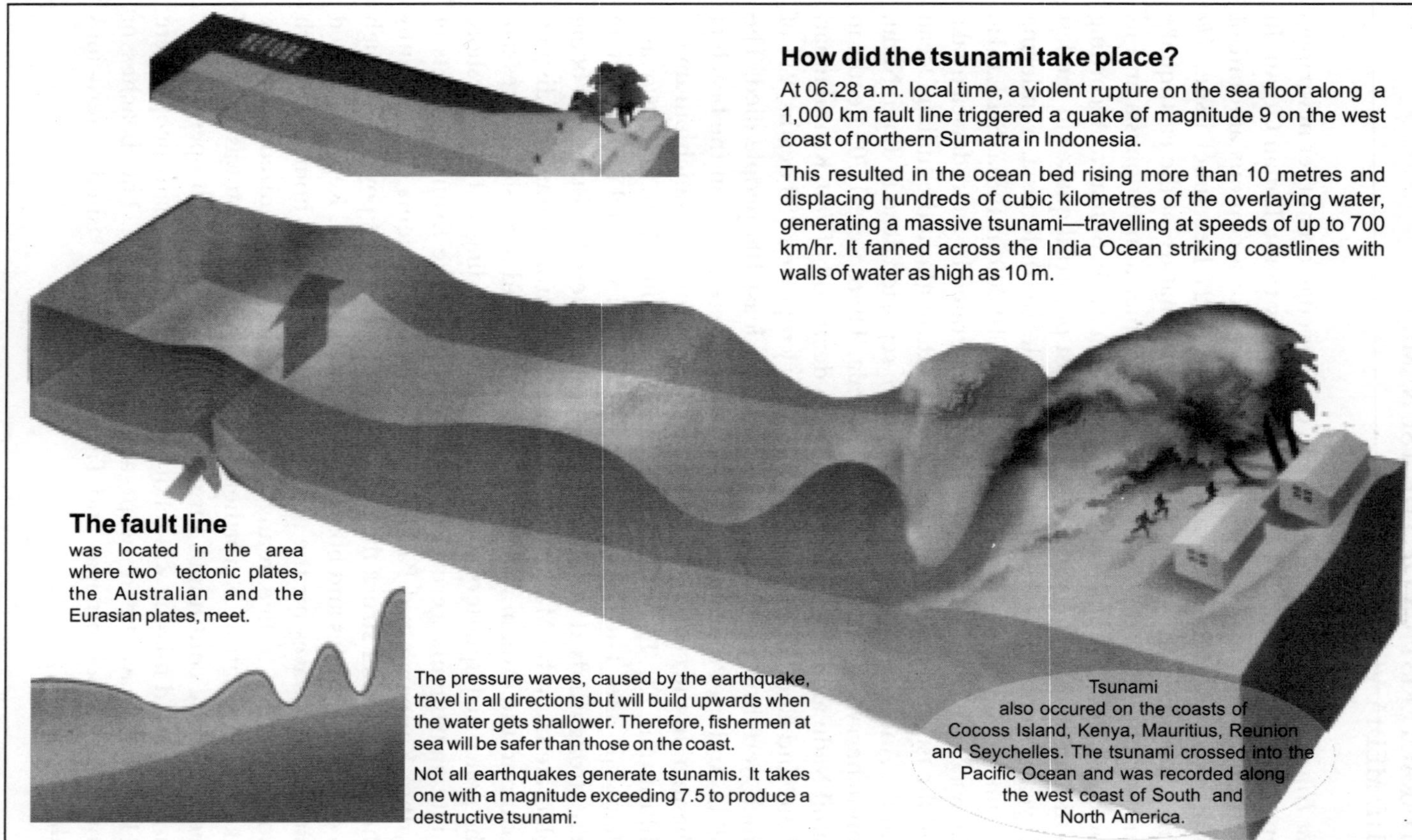

Fig.14 The cause of tsunami

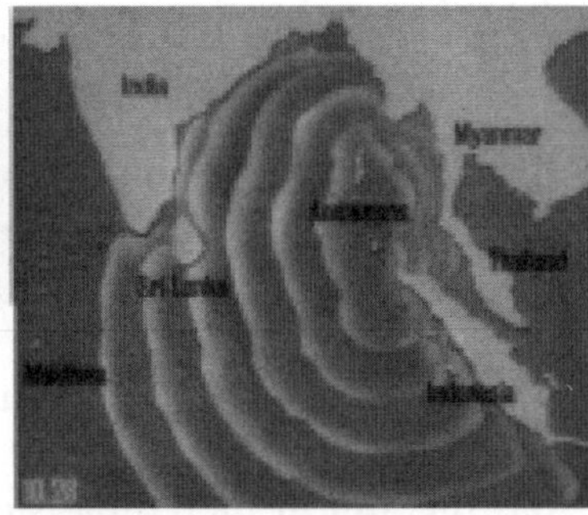

The Big Ones

April 1, 1946 : Aleutian earthquake and tsunami

Generated by a magnitude 7.8 quake near Unimak Island in Alaska's Aleutian Island Chain, this tsunami of 35 metres reached Hawaii killing 159 people.

November 4, 1952: Kamchatka earthquake and tsunami

A magnitude 8.2 quake off the coast of Kamchatka Peninsula generated this Pacific-wide tsunami. The height of the waves touched 15 m.

March 9, 1957: Aleutian earthquake and tsunami

An 8.3 magnitude quake generated a Pacific-wide tsunami causing extensive destruction of property in Hawaiian Islands. The waves reached a height of 16 m.

May 22, 1960: Chilean earthquake and tsunami

The largest earthquake (magnitude 8.6) of the 20th century off the coast of south central Chile generated a Pacific-wide tsunami and killed 2,300 people in Chile alone. Total damage in various countries was estimated at more than $500 million.

March 28, 1964: Alaska earthquake and tsunami

An earthquake of magnitude 8.4 struck an area extending from Valdez to the Trinity Islands, near the Gulf of Alaska, and caused areas to be lifted by as much as 15 m. It killed more than 120 people. Alaska's fishing industry and most seaport facilities were virtually destroyed.

Fig. 15 Major tsunamis in the twentieth century

The inhabitants of Besantnagar saw the first surge that rose over two metres. The sight of the second surge was yet more terrifying. It sent the fishermen and people scurrying for safety. The lack of any prior warning caused utter pandemonium. According to the media, the tsunami caused enormous damage to property from north Chennai's Ennore-Tiruvottojeer-Royapuram coast up to Kalapakkam, 60 km south of the city. Newspapers carried a large number of shocking and heart-rending stories about children trying to battle the tsunami. Between 27 December 2004 and 9 January 2005 *The Hindu* presented to the entire nation the human impact of the disaster. In its 27 December issue it told the almost incredible story of how a teenager wrestled against the powerful tides of the Bay of Bengal for two days before being washed ashore a snake-infested island. The article 'A girl's tryst with the tsunami' described how Piuli Paul 'miraculously escaped after being washed away by the big waves that hit the Chennai coast. "I was in the water for a long time. I was perhaps unconscious," she recalled. Piuli went to the Marine Beach on Sunday morning accompanied by her parents and grandparents. According to her "We were wetting our feet when a wave hit us. My *dida* (grandmother) fell on the sand. I went to help her and grabbed her bag. Others came to help too but the big wave struck them and swept us away." When she regained her consciousness she tried grabbing a bar to pull herself out of water. As she said, "I had lost all strength. One uncle then helped me come out of the water."' Piuli later found her parents and grandfather, but her grandmother Bela Paul (73) was killed by the wave. Equally horrifying is the account of the engineer whose wife was washed away by the wave before his eyes, despite his efforts to save her. The Cuddalore coast was especially badly hit. Calamity struck the coastal villages of the Cuddalore district at about 8.30 am and the waves hit the coastal areas at least five times within four hours. Hundreds of people were reported missing and several others injured when the waves rose from two to three metres, causing the tenements on the shore to collapse. Even seasoned fishermen did not dare to swim against the power of the waves; women and children were helpless victims of the fury of the seas. The worst affected coastal villages were Mallavadu, Suba, Uppalavadi, Thallangudi, Thevanampattinam, Sonamkuppam, Samiarpettai, Pudhukuppam, Muzhukkuthurai, MGR Thittu, Killai Periakuppam and Chinnur. According to the district collector there were 218 deaths, with several persons missing. Over 400 fishermen and their family members were reported killed and thousands of people rendered homeless as waves hit the coastal areas of Kanyakumari district. Seawater entered all the coastal villages of Kadiyapattinam and Colachel. The 3 km road between Colachel and Vaniyakudi beach and Sothavilai beach was completely flooded. The district administration declared the death of 270 persons in Colachel, 10 in south Vilai, two in Pakam, 11 in Muttom, two in Kanyakumari, four in Manakudi, 13 in both Azikkal and Manavalakurchi and 75 in Asaripallam.

The media reported that according to the locals at Nagore, Velankanni and Nagapattinam, the death toll exceeded 3,000 and there were mass burials at each of these places. Around 75 relief centres became operational in the affected areas and more than 40,000 people had to be evacuated. Nagapattinam became a watery grave for the fishermen and fisherwomen for whom the sea was a means of livelihood. For instance, the fishing colony at Seruthur, about 1 km from the Velankannai shrine was completely wiped out. On the day after the tsunami the number of people still alive in similar colonies nearby dropped from 1,800 to 1,100. 230 fishing boats, including those built of expensive fibreglass, and 30 catamarans were damaged, apart from the large number that had been washed away. Each fibreglass boat along with its fishing net was estimated to cost over Rs 100,000. The fishermen who customarily borrowed money from private lenders at high interest rates for purchasing equipment suddenly found that they had lost not just their families but also their livelihood. The media focused on the tragic predicament of Kalyani, a fisherwoman at Nochikuppam. The lone survivor in a family of four, the media used her desperate cry 'Why has the sea left me alone?' as a paradigm of human suffering caused by the disaster. Most families in the area had suffered similar tragedies.

Although the slum dwellers could not ultimately escape the fury of the waves, they put up a brave stand and some of them even staunchly refused to seek refuge elsewhere. In Kanchipuram the death toll rose to 82; of them around 42 people died when water entered two churches during the Sunday mass. The community centre behind St Anthony's Church and the children's playground, were completely submerged and the receding water left a trail of destroyed buildings. At Oyyali Kuppam, little remained of the buildings constructed by the administration in the past couple of decades. Three children were reportedly stranded on a small island in the village, but despite a frantic search using a helicopter only the bodies of several women were found. In the nearby village of Pudupattinam, fibreglass boats, catamarans, fishing nets, huts and concrete houses were all reduced to rubble. The number of human casualties remains unconfirmed.

Damage assessment continued for a while after the tsunami. For example, on 29 December 2004 there was a report that 13 more bodies had been recovered from the coastal areas, raising Chennai's death toll to 173. On the same day 500 more bodies were found in the district of Nagapattinam, raising the death toll in the coastal district to 2,500.

Heritage buildings like the Shore Temple at Mamallapuram in Chennai, were not spared by the tsunami. The waves entered the temple after demolishing the barriers around it and although the Shore Temple was not physically damaged, there were fears that the waves may have dislocated the foundation of the *balipectam* (sacrificial altar) in front of it. The Archaeological Survey of India (ASI), Chennai Circle has made significant observations about the architectural structures that remained submerged within and beyond the temple. There were disturbing rumours that the tsunami had hit the Madras Atomic Power Station (MAPS), Kalpakkam, which housed two pressurized heavy water reactors. While the waves indeed hit the township about 6 km from the plant, they did not hit the plant itself. The Chairman of the Atomic Energy Commission publicly declared that the second reactor was quickly shut down and the first had not been operational. According to him, the rise in the sea level caused a corresponding rise in the pump house, which kept drawing water from the sea for condenser cooling. This was detected by the operator in the control room, who decided to shut down the reactor; there were no damages to the turbine or reactor building. The Chairman was categorical in his assurance that there had been 'no radiation leak' from the reactor. Although the construction site of the 500 MW Prototype Fast Breeder Reactor (PFBR) was flooded, the water could be pumped out.

Several other parts of India were also affected by the tsunami. Kerala, for instance, was ravaged by the flood. On 27 December, *The Hindu* described the tsunami waves from the Arabian Sea that wreaked havoc in the coastal fishing hamlets in the southern part of Kerala. It is believed that 100 people were killed, the toll being the highest with at least 86 deaths in the Kollam district, 21 deaths in the Alappuzha district and perhaps two in the Ernakulam district. The coastal villages in the Malabar region suffered a little less. At places like Puthiyappa and Kolhi at Beypore in Kozhikode district, fishing canoes were battered by the waves. In Muzhipplilalgadi village in Kannur district, a fisherman gave his life to save a canoe from the violent waves. There were about 10,000 evacuees from the affected villages of Alapuzha district. On 28 December *The Hindu* carried a tragic commentary on the devastation in Kerala. The people of Azheekal in Kollam were panic-stricken at the rising water level; many of the women in the district died trying to save their children; a physically disabled man who narrowly escaped being drowned by the sea waves watched helplessly as his two children were blown away by a furious gale. Much like their counterparts in Tamil Nadu, many of the fishermen in Kerala lost their families. The 150-year old Paschimeswara Temple of Durga and Subramanian belonging to Arya Swapam was washed away leaving only the flag mast. The Alappad village in the Kollam district and the Arattupuzha village in Alapuzha district were devastated by the tsunami and the total death count rose to 170. This prompted the government to undertake rehabilitation work in earnest.

For the Andaman and Nicobar Islands the 2004 tsunami was the first in more than 60 years. The earlier tsunamis of 1941 and 1881 were triggered by seismic activity in the region. These islands were among the worst hit by the Sumatra earthquake, followed by the tsunami and about 18 aftershocks. According to the administration, tremors were felt there at 1.30 am, 6.25 am and 8.15 pm. The Andaman and Nicobar Islands are prone to seismic disturbances as they are close to the zone where the section of the earth's crust on which India and Australia lie slides below the section of the Asian land mass. These two sections converge at the rate of 5.4 cm a year. As the sections slowly slide past one another, *displacements* rather than tension occur at different points, which are manifested through earthquakes. The undersea earthquake off Sumatra might have occurred in the vicinity of these islands. One can rely on the US Geological Survey (USGS) data. According to data on the USGS website, the earthquake measured more than 7 on the Richter scale. It showed that the earthquakes on these islands had occurred up to 9° northwards and 2° westwards of the one in Sumatra; thus tremors from the Sumatra earthquake might have triggered earthquakes in and around these two islands. One may recall the earthquakes of comparable intensity in 1881 and 1941 that caused tsunamis.

Within 72 hours of the incident the death toll was around 10,000 and 25,000 had been rendered homeless. The worst affected populations were on the south-eastern side of the islands. In the first few days more than 60,000 people in the Little Andamans, Car Nicobar, Nancowrie islands and Campbell Bay were completely cut off. The aboriginal and tribal populations felt the full force of the disaster.

In other states of India, particularly Karnataka and Andhra Pradesh, the situation was relatively better. Andhra Pradesh has a 1,000 km-long coastline inhabited by a few hundred fishermen. The tsunami struck the coast at around 8.45 am and rose 2-5 metres high, claiming about

84 lives. There were 32 deaths in the Prakasam district; on the Manginipudi beach 26 bathers were killed. Both west Godavari and Nellore had 11 deaths each; there were three deaths in east Godavari and one in the Guntur district. Although Visakhapatnam, Srikakulam and Tirupati felt minor tremors, there was hardly any major damage to structures. Similarly many sensitive structures of the Department of Space and Defence, situated fairly close to the sea, were spared any damage. However, water entered the launch pad of the SHAR centre, Sriharikota. The deaths in Karnataka were by and large restricted to tourists visiting the state. The figure reportedly rose to 49, with more than double the number missing. At Velankanni on the Karnataka beach, around 25 people were washed away.

Although the Indian coast is 2,000 km away from the Sumatra coast, the tsunami in the Indian Ocean, according to initial reports, resulted in heavy casualties across the coastal districts in Sri Lanka. There were more than 4,500 deaths and at least 22,000 people were rendered homeless. The coastal belt affected by the tsunami extended from the northern Jaffna district to the western Colombo district. The entire coastal population felt the tremors around 6.30 am, as did those in some other parts of the island, including the Central Hill districts. The northern part of the Jaffna peninsula was as severely affected as many parts of India.

Later, the death toll rose to almost 11,500 persons and showed signs of increasing by a few thousand more as the waters began to recede from the coast. A total of 10,029 civilians were killed and 48,000 left homeless on a day when they were preparing for Christmas festivities. North-eastern Sri Lanka was very badly affected by the tsunami. Two to three days later, the official death toll rose to nearly 22,493 and a million Sri Lankans were left homeless. Corpses were found in rivers, lagoons and mounds of debris. The quality of water deteriorated rapidly, causing concerns about the outbreak of an epidemic. In the southern city of Galle, nearly all homes within about 1 km of the coast were wholly or partly destroyed. With evacuees crowded in churches, mosques and Buddhist temples the city experienced severe shortages of food. Several foreign tourists were killed by the tsunami. The Disaster Management Unit (DMU) based at the President's Office continued to provide official figures and took measures for relief and rehabilitation. At Panadura, around 30 km south of Colombo, the entire community participated in relief activities.

Hundreds of Sri Lankan doctors working in the UK volunteered their services. The hospital morgues at Panadura were filled to overflowing and the hospital staff were at a loss how to deal with the overwhelming number of corpses. World Vision (a Sri Lankan organization) expressed concerns about the possible spread of waterborne diseases and Oxfam sent their flood experts to Sri Lanka. The British media reflected on the inadequate support systems in Commonwealth countries such as India, Sri Lanka, Maldives, Malaysia and Bangladesh, all of which were hit by massive tidal waves. Soon after the tsunami hit the shores of East Africa on Sunday, the death toll in Somalia rose to around 110. Entire villages were swept away, widespread damage was caused and 55 people were killed in the island nation of the Maldives in the Indian Ocean.

The devastation caused by the tsunami was so great that its economic and commercial impact could barely be assessed accurately. The damage far exceeded that caused by other natural calamities. Along with local infrastructure, several heritage sites, tourist attractions and structures of great historical and cultural value were destroyed. The residential outfits of the Kalpakkam Plant were damaged and commercial boats suffered great damage too. As there was virtually no warning or pre-tsunami phase, the tsunami itself struck people as being especially destructive and relentless. The support and relief efforts that began soon after the disaster foreshadowed broader measures for tsunami mitigation. The media questioned the efficacy of safeguards against tsunamis and the reassessment of existing public policies began.

POST-TSUNAMI: ISSUES VIS-À-VIS POLICY IMPLICATIONS

The day after the tsunami, most newspapers merely reported the event, but did not publish analytical articles about the disaster and issues of preparedness and disaster management. The suddenness of the tsunami took almost everyone by surprise. Indian forecast mechanisms failed to warn that the tsunami would soon hit the coast. The prior external manifestation of a tsunami is a deceptively small elevation of the sea level, following an earthquake. If the occurrence of the earthquake and the rise in the sea level go unnoticed, it is impossible to be forewarned of the tsunami. The meteorological establishment was harshly criticized for failing to alert the general public—apparently forecasters were not adequately equipped to issue a general warning. Although the tsunami is indeed a rare disaster, Indian forecasting systems need to be critically reviewed. People in the Andaman and Nicobar Islands did not anticipate the tsunami even though there may have been hints of disturbance. While India certainly relies on state-of-the-art equipment, evidently the latter was incapable of predicting a disaster of the tsunami's magnitude. Commentators and the media asserted that India lacked a definite statutory obligation for issuing a public notice/ forewarning about natural disasters. Some were skeptical how effective an advance warning may have been, given that most were people were asleep at the time. A newspaper editorial commented on the helplessness of meteorologists and specialists as they struggled to come to terms with the 'realization that formidable scientific capabilities could not generate so much as a hint that a killer wave might follow the first and the strongest of the 14 earthquakes in the region'.[2] Although several organizations, especially NGOs, came forward to help, the relief effort did not quite rise to the occasion immediately. The Association for India's Development (AID) was among the few NGOs that ensured that supplies reached even the poorest; Indians for Collective Action along with AID were keen to expand their activities from relief to rehabilitation and corporate support offered some relief to the 35,000 families in Nagapattinam and Cuddalore. In some cases, volunteers offering assistance and support to the fisherfolk in different villages even provided them with their earlier means of livelihood, such as fishing boats, catamarans, etc. The IT organizations in the corporate sector readily offered support. The general public seems quite reconciled to the official view that the prior allocation of funds for tsunamis is hardly expected as it is such a rare event. Strangely, no skepticism was evinced, even by the media, that the occurrence of tsunamis was not included in the category of Disastrous Events for Prior Planning despite earlier recommendations for the same during the IDNDR. Although there were reservations about developing countries accepting aid from developed countries, public opinion insisted on the establishment of proper infrastructure to defend against future tsunamis in the Indian Ocean. The tsunami warning system must be able to detect deep sea movements and a network for sharing information with other countries in South East Asia needs to be developed.

The President of India, Dr A P J Abdul Kalam, shared his thoughts and ideas with the university community and called for a tsunami warning system along the entire Indian coastline, modeled on the existing system for the 27 Pacific nations.[3] The paucity of resources led to an inaccurate and limited analysis, resulting in widespread deliberations on the policy on disaster management. A few experts from abroad, such as Prof. T N Murthy of the International Tsunami Society based at Canada, tried to persuade the establishment to adopt all possible methods for forecasting future tsunamis.[4] Quite naturally, there were debates and controversies about the failures of disaster management systems. It is increasingly felt that blaming functionaries is of little or no avail. There is a growing tendency to look for technological solutions for disasters rather than possible non-structural alternatives. Corporate India may not yet have appealed to the government for economic policy to include disaster management, but a meaningful balance between economic concerns and disaster preparedness must be achieved on the policy agenda. While improved warning, evacuation and sheltering systems are the chief constituents of Central Government policies, the communitarian perspective ought to emphasize viable choices from among available alternatives.

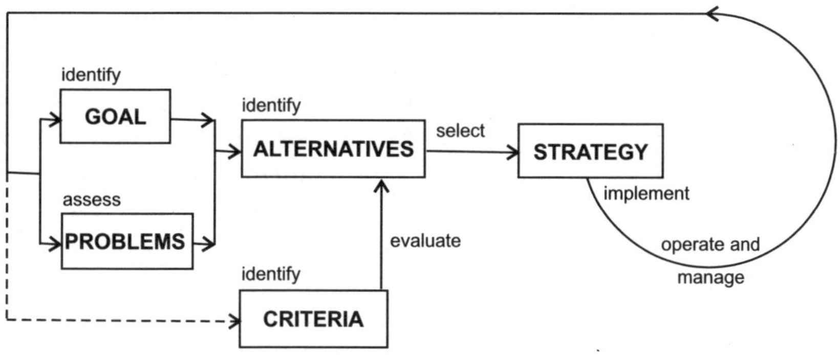

Fig. 16 Policy implementation process

REVISITING UNDERLYING ISSUES

None of the countries in South East Asia are immune to natural disasters. Some of them—particularly Vietnam, Lao PDR, Cambodia, Korea, Taiwan—suffer primarily from water-related hazards. Neither seismicity nor tsunamicity seem to occur frequently in Japan, China and India. Indonesia however, is an exception. International organizations such as the ESCAP and UNCRD frequently deliberate on natural disasters, which are matters of common concern. Media reports after the recent tsunami drew attention to coastal vulnerability—Sri Lanka's coast was probably the most vulnerable, despite the fact that Sumatra was probably worst damaged. Although the Philippines escaped the ravages of the recent tsunami, its earlier history shows that its islands, particularly the Mindoro, have experienced the full fury of tsunamis. Despite much deliberation and a few case studies, there is no detailed analysis of disaster management in some of the South East Asian countries. There are still inhibitions about and a limited tradition of social support for research in the wake of natural disasters. Research and planning are accorded low priority. The affected countries are yet to investigate relevant phenomena from the past that have escaped detection. Such comparative analyses would facilitate a deeper understanding of contemporary natural disasters. The recent tsunami stimulated the media to delve into past records of tsunamis and make comparisons, but these exercises have been largely *descriptive* and have lacked focus on the science of such a phenomenon. Undeniably there are now forms of communication and dissemination other than newspapers, the radio and television. IT is the latest buzzword and reports indicate that Internet connections could be crucial for rescue and evacuation operations. Internet enthusiasts in South East Asian countries created the South East Asia Earthquake and Tsunami (SEA-EAT) web blog. This soon became a major online platform for the community to share information and contact details. It was a novel non-official venture helping not merely the survivors of the disaster but also desperate relatives searching for the missing members of their families. Apart from information on missing persons the SEA-EAT blog also provided the latest updates on relief operations in countries like Thailand, Sri Lanka, Indonesia and India. The establishment of virtual rescue centres, easily accessible to the public, is a landmark initiative. Blogging has become a powerful tool for the dissemination of information. After the disaster, it proved to be an excellent alternative to the overloaded official helplines that inevitably provide only partial information. Of course, people also extensively used conventional media outlets (such as online versions of newspapers) to inquire about their relatives and friends. The websites of certain widely trusted organizations like the BBC were flooded with information.

Despite the presence of emergency/crisis management groups in different countries, public opinion seems to be steadily building in favour of a comprehensive approach that includes security, safety and protection against natural hazards. The activities of private agencies and NGOs during post-disaster phases, emergencies and relief operations did not lack focus, but disaster management is generally the responsibility of the police, fire services and hospitals, and on critical occasions, the army. There is a strong need for regional emergency planning coupled with services provided by the local authority. As demonstrated in Indonesia and the Andaman and Nicobar Islands, the intervention of defence organizations yields benefits. In the aftermath of the disaster, self-help, mutual support and community collaboration was widely publicized in the mass media. Of course, external observers had much to say about the apparent lack of attention to flaws and shortcomings in the planning/preparation for disasters. The media carried a large number of interviews with survivors, evacuees and analysts/ commentators. They also cited the official response to the disaster and suggested means for tackling it. It was evident that recovery operations were more efficient if members of a particular community had prior experience working as volunteers for relief programmes. There was a universal feeling that services were inadequate for a major emergency caused by a tsunami. Very little was said about the lack of professionalism needed to counter difficulties in the aftermath of the tsunami. The emergence of other issues and the obvious absence of a comprehensive plan of action ensured that grievances about the inadequacy of emergency management were not forgotten. The media was relentless in its inquiry into the economics of the disaster and possible future disasters. For the first time, the effect of a disaster on the sensex was critically examined. Criticism of the poor forecasting methods, accusations of the establishment for the same, etc. caused concerns about falling share prices and financial reversals. Such analyses underscore the changes in the image and power of the mass media. The findings of the media ought to be taken into account while formulating policy decisions. Limitations of policies on disaster management have a domino effect on other activities. These limitations might cause an adverse shift in structural adjustments, which in turn has an effect on global financial trading, and ultimately, even on global policy.

ANALYTICAL APPROACH TO THE RECENT TSUNAMI

It is deeply encouraging that the media went far beyond its predictable role of reporting of the tsunami's immensity and the resultant losses. Not just the daily media, but even weekly and fortnightly periodicals carried knowledgeable expositions on the tsunami. The gripping portrayals of affected regions did not degenerate into vulgar sensationalism, but had an objective and scientific flavour to them. They were apt responses to readers' curiosity about natural hazards. The tsunami drove home the human consequences of natural disasters; people realized that understanding the symbiosis between human beings and their environment was of paramount importance.

Risk perception, assessment, cognition, acceptance, etc. are often discussed in detail by the media and are probably highlighted as a result of socio-cultural psychological processes. An analytical approach is therefore an absolute necessity. The central question is how to formulate and interpret the situational dynamics of the phenomenon, let alone attempting to offer a solution. The process must be carried out step by step because of the periodic emergence of non-linearities in different areas. But first one should have a systemic view about natural hazards. Such a system must have certain components (which are called subsystems). There are many available schemata representing the totality of such a system having interactive subsystems—they are often referred to as natural disaster cycles. In order to identify the subsystems one needs to examine the contexts of the natural hazard, which depend on natural processes, population strength, etc. The easily identifiable components are (a) physical elements, (b) human elements (c) adaptability to hazards and (d) outcomes (loss). How are these elements to be measured? Can these elements be mapped into yet broader categories? If so, can physical elements be associated with risks, human elements with vulnerability, the adaptation to hazards with human responses and outcomes with costs? Such mappings are obviously related to the total capabilities and resilience of certain sections of society. The system, however, is not static. The degree of seismicity and tsunamicity in a region may change over time. There may also be changes due to population explosions or processes like migration or changes of habitat. Similarly, the adaptation to hazards may be affected by the development of communication technologies (remote sensing, satellite imagery, etc.).

Such exercises ought to focus on the multi-layered character of natural hazards. The extreme vulnerability of certain locales is exposed during natural disasters. In order to facilitate a closer and more exhaustive study of natural hazards, the broad notion of the 'context' of a disaster must be broken into certain subsets—such as spatio-temporal, institutional, environmental, politico-economical, socio-cultural and other factors—and each treated as a separate subject of study. Are these subsets interchangeable? If contexts vary does the definition of their constituent elements pose a dilemma? One must remember however that the whole is greater than the sum of its parts and an analysis of the components of a natural disaster should not blind one to the larger picture presented by such hazards. Simply because tsunamis have rarely occurred since 1881, one cannot neglect the danger they pose or underestimate the extent of the damage they could cause. The horrifying state of the Car-Nicobar exposed the proneness of the area to disasters. Governmental intervention through the defence establishment was an unusual deviation from the conventional policy of providing limited assistance to the affected inhabitants of the island. A great advantage to the aid-givers was the indigenous population's ethnic familiarity with seismic tremors and the effects of disasters. An intrinsic conditioning to natural hazards enabled them to cope with and overcome the havoc wrought by the tsunami. The disaster had far-reaching implications for future disaster management policies which ought to take into account both the 'physical' and economic consequences of the event and the response of the public.

CONCLUDING REMARKS

The variety of media and the rapid progress of communication technology have lent a new air of immediacy to news reportage. After the tsunami struck in December 2004, the media have been most courageous, revealing candidly the state of aid and support. The report of Action Aid International (AAI) bluntly revealed that two-thirds of the donations pledged by aid-giving nations had failed to materialize. Greater accountability had been expected of the richer nations and donor countries. There is a disturbing difference between the promises and actual provision of aid by donors. Investigations show that only about 60 per cent of aid actually reaches recipients. The failure to provide aid to the poorest countries is a grim reality. The large amounts of miscellaneous expenses by donor countries have greatly deflated the value of the aid. The AAI report has criticized the G-7 countries that have not delivered the aid they pledged. For instance, Australia has given 7 per cent of the money it committed to the tsunami emergency relief and reconstruction, France has given 13 per cent, Germany 15 per cent and the Netherlands 16 per cent. The US and the European Commission have donated only about 38 per cent of their pledge and Canada has given only 20 per cent, Norway 46 per cent and Italy 59 per cent. AAI feels that the flow of aid should be more transparent; there should be greater clarity of policies and criteria for aid and support. Over decades South East Asian countries have found themselves at the receiving end of the whims of global organizations. Although natural disasters in the region share some common features with those in other parts of the world, they are qualitatively different. While some South East Asian nations are affected by slow onset disasters, others experience (albeit more rarely) rapid onset disasters. All these factors taken together present a cogent case for global support to assume a truly scientific, transparent and timebound character, sensitive to the requirements of the affected countries. Given the susceptibility of South East Asia to natural disasters, a consortium ought to be created to facilitate the process of give and take with readily extendable and acceptable terms.

The clauses of the Tampere Declaration on Disaster Communications (1991) are yet to be followed up in a comprehensive and well-coordinated manner although the Annenberg Washington Programme is an important step in the right direction. This declaration generated many follow-up programmes during the IDNDR. Though in 2005 the WCDR embarked upon similar ventures, it focused primarily on natural disasters in and around the Indian Ocean. Even the Asian summit attached great importance to the vulnerability of the Indian Ocean to natural hazards. Recent developments in communication technology have ensured that the lack of timely transmission of information can no longer be a pretext for inadequate preparatory measures.

NOTES

1. The Annenberg Washington Programme/International Disaster Communications Project, *The Tampere Declaration on Disaster Communications*, Communications Policy Studies, Northwestern University, Chicago, 1991.
2. *The Hindu*, 27 December 2004.
3. Further insights are provided by President Kalam's excellent article, 'Natural disaster needs human innovations', *The Week*, June 2005, pp.32-5.
4. Radhika, V, and Ali, Farwa Imam, 'Track a tsunami', *The Week*, 9 January 2005, pp.36-7.

Chapter 3

A Résumé of Global Perspectives

INTRODUCTORY REMARKS

The IDNDR generated a large number of activities and programmes. Although the structured and well-planned disaster relief and management programmes began during the IDNDR, a few distinctive endeavours may be seen as forerunners of the IDNDR, e.g. the United States Decade for Natural Disaster Reduction (USDNDR). A few structured programmes were also undertaken in some developing countries. Notable among these was the 1987 endeavour by the South Asian Association for Regional Cooperation (SAARC) countries. The third SAARC summit held in Kathmandu, Nepal, in November 1987 deliberated on the 'impact that the environmental degradation including extensive destruction of forests and natural disasters such as floods, droughts, landslides, cyclones and tidal waves have had on the development process and prospects in the region and immense human suffering caused by them'. The heads of state of the participating countries were concerned with the threats posed by global climatic change, such as the rise in the sea level and its effects on South Asian countries. An expert group was set up to conduct detailed studies on the matter. The group presented its report at the tenth session at Kathmandu from 13-15 July 1988. The report seriously considered the SAARC countries' common concerns about natural disasters and the environment, and sought to formulate and work out strategies for disaster mitigation. Sustainable development by the year 2000 was certainly a vital goal for the SAARC. The geographical, socio-economic and regional imbalances in South Asia were keenly felt and measures to deal with environmental and natural disaster management were discussed.

The USDNDR was an important precursor of the IDNDR. It was launched in 1987 with an excellent document entitled *Comforting*

Natural Disasters: An International Decade for Natural Hazard Reduction (IDNHR) and a supplementary publication, *Reducing Disaster's Toll.* Both publications were prepared under the leadership of Prof. Frank Press and though they dealt chiefly with natural disasters in the USA, many of the issues discussed have global implications as hazard reduction processes the world over have certain common features. Therefore general comments and observations are relevant to non-American contexts. As an initiative involving a wide range of countries, the IDNDR drew upon the broader observations of the USDNDR. Indeed, the framework and documents of the USDNDR anticipated and to some extent established the rationale of the IDNDR. The second publication mentions that among natural hazards, tsunamis cause the most significant damage and loss in many parts of the world. The brilliance of the publication lies in its identification of key issues for the decade, including fundamental research on natural disasters.

GLOBAL PERSPECTIVES

The global attention that the 2004 tsunami commanded reinforces the claim that a natural disaster is certainly a global phenomenon. The relief and aid provided on earlier occasions were not comparable to the largesse shown by individuals, organizations and countries after the tsunami. Organized thinking in this area did not begin until the 1970s. The contemporary study of natural disasters is indebted to Prof. Frank Press of California Institute of Technology, USA. Also the President of the US Academy of Sciences, Prof. Press prevailed upon the UN to recognize the global issues surrounding natural disasters. Some of his remarks about the initiation of the IDNDR are even now considered extremely progressive and insightful. According to him, 'It is possible to reduce significantly human suffering and economic losses from natural disasters. This can be achieved by the appropriate use of the scientific and technological advances of recent years, by continued research to improve knowledge of natural disasters, by instituting programmes of training and sharing knowledge, and by raising the priority given to the mitigation of natural disasters by governments. These goals of replacing fatalistic attitudes by enlightened public policies that make use of advances in knowledge, are worthy of humane societies and their political leaders. They are also cost effective in that the economic saving from disaster mitigation programmes outweigh the costs.'[1]

Prof. Press emphasized the need for countries affected by natural disasters to share their experiences and ideas. Though couched in the customary UN semantics, the relevant UN resolution cannot simply be regarded as a catalogue of reasons for natural disasters. The global scenario was not particularly encouraging during the 1980s. Though *economic* globalization had made significant progress, the process of globalization had had very little positive effect in the area of natural disasters. Prof. Press insisted on detailed study and research in the field of natural disasters—thus the late 1980s and early 1990s witnessed efforts to dispel myths about natural disasters. The central question was: how should natural disaster reduction be stripped of ad hoc-ism and indifference? This challenged researchers, activists and non-political venturers—it appeared as though the changes needed were attitudinal, more than anything else. Many organizations thought it wise to abandon the older methods of handling disasters and gradually began to translate the tenets of the IDNDR into reality. Who motivated and pioneered efforts in this direction? How can one create a global culture for coping with natural disasters? What preceded the boom of the 1990s? Was it really a 'shining' decade? Was there any visible buoyancy during the decade that caused this boom in the activities and programmes of the IDNDR? Did it prompt the establishments in the South East Asian countries to do away with the prevailing obsolete methods for tackling natural disasters? Did the South East Asian countries perhaps seek an edge over developed European nations and America? The collaboration between the European Centre for Medium Range Weather Forecast (ECMRWF) of the UK and the National Centre for Medium Range Weather Forecasting (NCMRWF) of India was both fruitful and significant. The IDNDR launched a number of innovative programmes—appraisals of IDNDR projects must recognize the use of sophisticated technology that has done away with the earlier sense of complacency about natural disasters.

STRATEGIES

Computations are often made of the number of people who died, were injured, disabled, etc. but the number living in abject poverty is rarely calculated. At present, policies for social inclusion and accommodation in the post-disaster phase are lacking. This may be rectified by reforming disaster management itself. Given the rapid growth of the corporate sector, the role of the establishment needs to be redefined.

It is worth speculating how the corporate sector would evaluate the IDNDR and WCDR from the vantage of the present. While the corporate sector may argue for or against the contributions of these projects, the issue remains purely apolitical. Disaster mitigation must take into account universal human rights. Corporate performance and ethics need to ensure the distribution of justice. Perhaps the market mechanism can be studied for ways of providing opportunities for the affected individuals. Minimum guarantees for the latter need to be explored and people with past experience of dealing with natural disasters ought to be consulted. Comprehensive planning and considerate governance may help accomplish such goals. We need to act decisively in order to enhance opportunities for the community. In the final analysis, if the initial measures are not implemented neither the government nor the corporate sector can provide solutions for the destruction caused by natural disasters.

A FRESH LOOK AT THE 1990S

With advances in information and communication technology (ICT) it is far easier to review earlier natural disasters. As stated earlier, natural disasters have certain easily identifiable common features. The declarations of the IDNDR have been stated below for reference.

PART I: OBJECTIVE AND GOALS OF THE DECADE

The objective of the Decade (IDNDR) is to reduce through concerted actions, especially in developing countries, the loss of life and property and socio-economic disruption caused by natural disasters.

The goals of the decade are

a) To improve the capacity of countries to mitigate the effects of natural disasters;
b) To devise appropriate guidelines for applying existing scientific and technical knowledge;
c) To foster scientific and engineering endeavours aimed at closing gaps in knowledge;
d) To disseminate existing and new technical information;
e) To develop measures for the assessment, predication, prevention and mitigation of natural disasters through programmes of technical assistance and technology transfer, demonstration projects and public information and training.

PART II: POLICY AND STRATEGY

To reach the objective and goals of the IDNDR the governments concerned have been requested:

a) To formulate national disaster mitigation programmes and to integrate these in national development plans;
b) To participate during the decade in concerted international action and establish national committees in cooperation with the relevant scientific and technological communities;
c) To encourage local authorities to take appropriate steps to mobilize public and private support;
d) To increase public awareness of risk and its consequence;
e) To pay due attention to the health aspects of natural disasters;
f) To improve early access to emergency suppliers;
g) To keep the Secretary General of the United Nations informed of all developments, nationally and internationally.

KEY DATES

1984 Concept proposed by Dr Frank Press at the Eighth Conference on Earthquake Engineering
1987 Concept of the IDNDR approved by the UN (Res/42/169)
1988 UN Secretary General appoints International Ad hoc Group of Experts to recommend a framework of action (Res/43/202)
1989 Report submitted by International Ad hoc Group of Experts (DOC/44/322)
1989 UN General Assembly proclaims the IDNDR and adopts the International Framework of Action (Res/44/236)
1990 UN Secretary General appoints Director of the IDNDR Secretariat.
1990 Secretary General announces members of the Scientific and Technical Committee (STC).
1991 First meeting of STC is convened in Bonn.

IT played an important role in shaping the different phases of the IDNDR. As is customary with UN declarations, there were mid-term, penultimate and final reviews and then an overall reappraisal of the IDNDR deliberations. Several countries, even those not readily susceptible to natural disasters, have been proactive enough to go well beyond what was expected of them by the tenets of the IDNDR. India was among the few countries in the developing world that did not rise enthusiastically to the occasion. The IDNDR earnestly sought partnerships but unfortunately, except for the ESCAP which took a leading role in the development of the region, willing partnerships have not been easily forged in South Asia. The much talked of workflow software platform does not seem to have materialized and the access to talent in different parts of South East Asia did not work out as desired. Perhaps the current social dynamics makes it necessary to develop workflow software. More web browsers and servers are also needed. A virtual global framework is needed to ensure easy accessibility.

The IDNDR has transformed the image of South Asian countries as active contributors to a sound initiative for disaster management. The IDNDR allowed the benefits of successful disaster management to reach the grass-roots level in some South East Asian countries. The efforts of the 'intellectual' community were instrumental in making valuable resources available in the new millennium. Despite the IDNDR agenda, there still remains some general cynicism about the malfunctioning of forecast and warning systems. At deliberations following the IDNDR at Yokohama and elsewhere, it was felt that the media should not be completely relied upon for an accurate reconstruction of post-disaster realities.

It is unwise to adopt a reductive view of South East Asian countries as they exhibit great multicultural, multi-religious and multi-ethnic diversity and even significant differences in entrepreneurial spirit. The enormity of natural disasters poses fundamental challenges. Countries' internal fragility is most evident in the immediate aftermath of a disaster—it becomes imperative then to handle the situation deftly and sensitively. A new framework ought to assure the restoration of normalcy, exploring new methods of restoration, rehabilitation and renovation. It is important to note that the IDNDR was not a product of empty rhetoric, as some seem to claim. In practice the IDNDR conformed to the ethics of the international community. Mobilization by UN organizations and other agencies was remarkable although some countries like India did not seem particularly enthusiastic about the IDNDR.

It is important to refer to the South Asian Regional Report that was presented at the world conference of IDNDR at Yokohama, Japan from 23-27 May 1994. This report essentially grew out of the proceedings of the SAARC Workshop on Natural Disaster Reduction, New Delhi, 30 March-2 April 1994. It provided an overview of natural disasters and the status of disaster management in SAARC countries. It took note of the third SAARC summit held in Kathmandu, Nepal followed by the sixth SAARC summit held at Colombo, Sri Lanka. At the Colombo meeting, a few areas for regional study and cooperation were identified, such as the networking of institutions of disaster management, relief and assistance mechanisms and the development of warning systems. The need for disaster victims to share their experiences and the need for a separate programme related to drought management were also emphasized. Prior to the Yokohama conference, there was a workshop sponsored by the SAARC on disaster preparedness and management in Dhaka from 17-19 October 1993. Geographically, the SAARC region comprises one-third of the population of Asia and one-tenth of the land mass of Asia. The problems of underdevelopment and the range of hazards experienced by the region are fairly well known. Natural calamities are basically geological and climatic in origin and nature. Bangladesh has the most disturbing record of cyclones and tidal surges; earthquakes have occurred in most of the SAARC countries except Sri Lanka and Maldives; drought affects all the countries in this region except Bhutan and Maldives. Pakistan, Nepal and Sri Lanka suffer extensive damages due to floods. Landslides occur in Pakistan, India, Nepal and Bhutan. Almost all types of natural disasters occur in India. The recent tsunami affected all the SAARC countries with coasts adjoining the Indian Ocean.

The SAARC countries have for the most part adhered to the priorities identified by the IDNDR, and have undertaken risk and vulnerability analysis. Policymakers realized that efforts on risk assessment need to be updated. Vulnerability analysis poses a different challenge altogether, mainly because of socio-economic factors. According to the report there exists 'differential vulnerability' in the broader context of disaster reduction and mitigation measures. These types of vulnerability, therefore, need to be reduced in phases. The development strategy has to adapt itself to what may be called the 'heightened vulnerability' of socio-economically disadvantaged segments. Of course, coastal vulnerability heads the agenda of combating natural disasters.

WCDR

The key dates mentioned earlier illustrate a significant pattern that has emerged after the declaration of the IDNDR. With unfailing regularity decisions are made and projects undertaken, adhering to international standards. The mid-term appraisal of the IDNDR was made at the World Conference on Natural Disaster Reduction at Yokohama from 23-27 May

1994. Even though there had been earlier conferences on the IDNDR, this conference took stock of events till the middle of the decade more effectively. Eminent personalities, including Boutros Boutros Ghali the then UN Secretary General and practitioners and policymakers from disaster-prone countries participating in the conference drew lessons from the past in order to help shape the future of disaster management.

During the IDNDR, humanitarian considerations assumed critical importance for the first time. The deliberations at Yokohama recognized the effect of natural disasters on human welfare; development programmes would thus necessarily address this issue. Another significant result of the conference was to review the promotion of culture of science and technology, effectively initiated during the launch of the IDNDR. The Yokohama conference aimed to achieve sustainable development worldwide by 2000. The details of the Yokohama Strategy are cited in the bulletin *Stop Disasters* (May 1994).

The next important landmark was the World Conference on Disaster Reduction at Kobe, Hyogo, Japan from 18-22 January 2005. It was a sequel to the UN Resolution 57/256 for a review of the Yokohama Strategy and Plan of Action for a safer world, guidelines for natural disaster prevention, preparedness and mitigation and its plan of action, which were reiterated in the UN Resolution 58/214. The Yokohama review was essentially a critique of the period 1994–present (then January 2005). As stated earlier all countries were supposed to achieve the various IDNDR targets by the year 2000.

Different nations and particularly the developing countries had certain expectations of the IDNDR. These expectations focused on promoting public awareness and sensitizing policymakers. It is rather unlikely that cutting-edge science and technology will be readily available to people in the developing world. Mobilization, campaigns and similar activities, backed by political will and NGO support, provided a positive response to natural disasters, even though new studies and publications could hardly be accessed by the community leaders. The IDNDR amassed public, governmental and political support for the cause of disaster reduction. Every new endeavour demonstrated the tremendous scope of evolving processes and programmes related to natural disaster management. The World Summit on Sustainable Development in Johannesburg added a new dimension to the existing system of linking natural disaster to 'development'. The two vital ingredients of the disaster framework being envisaged under the aegis of UN deserve a closer look. One ought to refer to two important UNDP articles, 'Living with Risk: A Global Review of Disaster Reduction

Initiatives' and 'Reducing Disaster Risk: A Challenge for Development', in *Know Risk* (2004). As their titles imply, these publications (and various CD-ROMS, auxiliaries, etc. released subsequently by the UNDP) bring to the fore a wealth of critically relevant data, country profiles and case studies of experiences *in situ* from more than 120 countries. These documents present a résumé of recorded disasters since 1994, along with their horrifying consequences. The authors of these publications and specialists belonging to leading organizations similar to the UN investigate the relation of disaster risk and vulnerability to development. There need to be continuous assessments of risk reduction vis-à-vis development. One can distill a pattern from the activities of the leading Asian countries engaged in such enterprises.

The concern for actual, potential and newly emerging risks is evident in the documents presented at the WCDR. In particular, there are references to the participation of international governmental and corporate organizations in countries like China, Philippines, Indonesia, Japan, Vietnam, Nepal and India. One should add that the Asian Disaster Reduction Centre was established as a specific response to the Yokohama Strategy, the ADPC having been established in Bangkok in 1985. A major focus area is the development of a disaster-related database. Although it is indisputable that there should be general awareness about the human impact of disasters, neither the IDNDR nor the WCDR provide any explicit details about how to respond to the needs of disaster victims. Also conspicuously absent is the emphasis on creating opportunities that have the potential to mitigate loss. The recent tsunami poses the critical question: is the decade of the WCDR to focus on *opportunities* rather than on *needs*?

The WCDR, though apparently an extension of the IDNDR, has probed more critical issues. Although the IDNDR was the foundation for later strategies such as the Yokohama Strategy, the WCDR certainly has an edge over the IDNDR. The pursuits of the IDNDR were not entirely compatible with the globalization process, which has now made perceptible inroads into the economics of countries that frequently experience natural disasters. The gamut of activities of the IDNDR were spread over a decade, but its deliberations with their markedly global focus and the lessons learnt from IDNDR programmes were concentrated into a single year (2005-6) by the WCDR. The escalation of disasters into global phenomena was amply demonstrated by the 2004 tsunami and the WCDR's recognition of the global impact of natural hazards and its condensation of earlier projects and principles of disaster risk management explains the following distinction: the IDNDR was called the decade of 'international disaster realities' while, thanks to the WCDR, 2000-10 is considered the decade of '*global* disaster realities'.

Conversely, one cannot ignore the effect of natural disasters on globalization. The IDNDR may not have considered an evolutionary approach, but non-linearities of natural disasters have continued to evolve. Such an approach may yield some mechanisms for policymakers to apply to disasters *in situ.* The culture of detailed analysis should rectify the present distortions in the formulation of policies and help policies to cope more efficiently with emergency management. Hopefully, unethical practices will also be avoided while dealing with natural disasters. If the perception of risks, vulnerability or hazards assumes a global dimension, there ought to be a marked difference in practices of combating natural disasters. The effects of natural hazards differ from one locale to another, and even the effects of the same disaster, e.g. a flood or an earthquake, in different locales could be remarkably different. These distinctions must be recognized. While the WCDR report analyses the various phases of natural disasters, the WCDR cannot magically redress the pitfalls of disaster management at the beginning of a new decade. People have barely recovered from the nightmare that was the tsunami. An urgent need is felt not simply to rehabilitate and rebuild, but to rebuild *differently* and to create the infrastructure for improved living conditions. On the whole, the WCDR document presents in unambiguous terms several strategic paradigm shifts, in some ways conceptually and qualitatively different from the rationale of the IDNDR.

WCDR: OBJECTIVES AND STRATEGIC GOALS:

1. OBJECTIVES

The World Conference on Disaster Reduction was convened as a result of the decision of the General Assembly, with five specific objectives:

(a) To conclude and report on the review of the Yokohama Strategy and its Plan of Action, with a view to updating the guiding framework on disaster reduction for the twenty-first century;
(b) To identify specific activities aimed at ensuring the implementation of relevant provisions of the Johannesburg Plan of Implementation of the World Summit on Sustainable Development on vulnerability, risk assessment and disaster management;
(c) To share good practices and lessons learned to enable further disaster reduction within the context of attaining sustainable

development and to identify gaps and challenges;

(d) To increase awareness of the importance of disaster reduction policies, thereby facilitating and promoting the implementation of those policies;

(e) To increase the reliability and availability of appropriate disaster-related information to the public and disaster management agencies in all regions, as set out in relevant provisions of the Johannesburg Plan of Implementation.

2. EXPECTED OUTCOME

Taking these objectives into account and drawing on the conclusions of the review of the Yokohama Strategy, states and other participants of the WCDR resolve to pursue the following expected outcome for the next ten years:

(a) The substantial reduction of disaster losses in lives and in the social, economic and environmental assets of communities and countries;

(b) The realization of this outcome will require the full commitment and involvement of all actors concerned, including governments, regional and international organizations, civil society including volunteers, the private sector and the scientific community.

3. STRATEGIC GOALS

To attain this expected outcome, the conference resolves to adopt the following strategic goals:

(a) The more effective integration of disaster risk considerations into sustainable development policies, planning and programming at all levels, with a special emphasis on disaster prevention, mitigation, preparedness and vulnerability reduction;

(b) The development and strengthening of institutions, mechanisms and capacities at all levels, in particular at the community level, that can systematically contribute to building resilience to hazards;

(c) The systematic incorporation of risk reduction approaches into the design and implementation of emergency preparedness, response and recovery programmes in the reconstruction of affected communities.

We now turn to another set of deliberations on the December 2004 tsunami in which several global players participated.

THE JAKARTA (ASEAN) SUMMIT[2]

Soon after the WCDR, the Emergency Tsunami Summit was convened by the members of the Association of South East Asian Nations (ASEAN). The miniature global summit held at Jakarta focused on the ASEAN countries affected by the tsunami and was especially significant because of the active participation of the UN Secretary General Kofi Annan. According to the UN, the events of 26 December 2004 were an 'unprecedented global catastrophe' which required an unprecedented global response. The summit essentially concentrated on long-term aid.

The other leading figures who participated in the summit were the Japanese Prime Minister, the Chinese Prime Minister, the Australian Prime Minister, the US Secretary of State and a score of other leaders from Asian countries, including 14 heads of countries and international organizations. The Indonesian President in his opening remarks stressed the urgent need for having a tsunami warning system in the Indian Ocean. The UN Secretary General made an impassioned appeal at this summit for support of US $979 billion for the ASEAN countries, taking account of the pledges already made. The World Health Organization (WHO) warned against the outbreak of infections, disease and the lack of availability of safe drinking water that might cause diseases like cholera and dysentry. According to reports Mr Annan and the WB president James Wolfensohn flew over the Sumatra island's west coast and then to the shattered coastal town of Meulaboh where there were piles of rubble six feet high and the authorities extracted 7,000 bodies from the debris. The media carried many reports of devastation in Malaysia, Thailand, Sri Lanka and the Andaman and Nicobar Islands, with the latest figures on fatalities. At their meeting, the ASEAN leaders resolved to initiate action-oriented programmes to strengthen emergency relief and rehabilitation in the aftermath of the tsunami. In particular, the declaration at this conference stated that 'the task requires resources which cannot be borne by any individual country'. The salient points of the agreement reached at Jakarta were classified under the following heads:

EMERGENCY RELIEF

(a) Mobilize additional resources to meet the emergency relief needs of victims.

(b) Request the United Nations to mobilize the international community to support the relief programmes in the affected countries. It welcomed the Flash Appeal by the UN. It will request the world body to appoint a Special Representative of the UN Secretary General for the above purpose; to convene an international pledging conference; and to explore the establishment of a 'standing agreement' at the global level for humanitarian relief efforts.

(c) Strengthen coordination and cooperation between national and international relief efforts to ensure effective and immediate distribution of assistance.

(d) Support the efforts of the affected countries to ensure an effective channeling and utilization of assistance from donors.

REHABILITATION AND RECONSTRUCTION

(a) Support and emphasize the importance of national rehabilitation and reconstruction programmes.

(b) Call on the international community, the World Bank and international financial institutions to provide funds for the viability and sustainability of those programmes.

(c) Establish a partnership, upon the request and the leadership of the country concerned, involving donor countries and regional as well as international financial institutions, to support the respective national programmes of the affected countries.

(d) Welcome the initiative of several countries on the moratorium of payments of the external debt of the affected countries to augment their capacity to carry out rehabilitation and reconstruction.

(e) Promote and encourage private sector participation for rehabilitation and reconstruction.

PREVENTION AND MITIGATION

(a) Support ASEAN's decision to establish regional mechanisms on disaster prevention and mitigation, among others, utilization of military and civilian personnel in disaster relief operations. Act as a regional instrument on disaster management.

(b) Establish a regional early warning system such as the Regional Tsunami Early Warning Centre for the Indian Ocean and South East Asia.

(c) Promote public education and awareness as well as community participation in disaster prevention and mitigation.

(d) Develop and promote national and regional human and institutional capacity, transfer of know-how, technology and scientific knowledge in building and managing a regional early warning system and disaster management through international cooperation and partnership.

The summit did not mince words about the initial flow of aid after the tsunami. The conference dwelt not only on the massive challenge posed by the event but also pleaded for the international community to assess what had gone wrong and examine the issue of long-term warning systems to be used for different disasters. To many world leaders the UN seems the most dependable body for organizing transparent and efficient programmes with long-term perspectives, such as having a warning system in the Indian Ocean. The Jakarta summit fuelled a spirit of camaraderie among ASEAN countries, especially those who had suffered major losses due to the tsunami. Both India and Pakistan sent naval ships to the Maldives to aid the rescue and relief operations. Apart from welcoming debt relief for countries hit by the 2004 tsunami, the participants of the Jakarta summit backed the decision to create an Indian Ocean early tsunami warning system which might save thousands of lives in the future. Kofi Annan appealed at the summit for US $977 million to cover basic humanitarian needs for an estimated 5 million people in the next six months. As reported by the *Times of India*, governments around the world pledged more than US $4 billion in aid, and private groups, corporations and individuals pledged another US $660 million.

In Sri Lanka the Indian Navy set up relief and medical camps and restored power, water and telephone lines. Considerable work was also done in Indonesia; two naval ships anchored off Meulaboh provided relief supplies. Of course, the Andaman and Nicobar Islands received extensive relief support as they are a part of India. The UN assessment of the number of deaths at the time was not very clear. It was felt that the true death toll may never be known, although according to the UN Emergency Relief Coordinator Jan Egeland, 'We are now counting new pledges by the hour. We are now between 1.1 billion and 2 billion.' The Jakarta summit was followed by a meeting in Phuket. The Phuket deliberations were largely about the installation a warning system in Indian Ocean and significant details about the mechanics of such a system were provided.

The IDNDR, the WCDR and the ASEAN summit were suitably global in character and their findings thus afford a *global* perspective of a regional disaster. Their deliberations emphasized two issues: the need for a warning system and the vulnerability of coasts adjoining the Indian Ocean. These have been discussed in detail in the following chapter. An evaluation of these issues must integrate global perspectives and local participation, without privileging any one element over the other.

NOTES

1. Press, Frank, 'Stimulating Progress in Science and Technology for Disaster Reduction', *Stop Disasters* No. 17, 1994, p.10.
2. For the facts and statements of this section I have used the 7 January 2005 issue of *The Times of India.*

Chapter 4

Coastal Vulnerability

The ocean continues to be an incredibly complex system. A rise in the sea level is considered a common phenomenon but it may not have been realized earlier that a deceptively small rise might herald a giant wave. 1988 was declared the International Year of the Ocean and the Intergovernmental Oceanographic Commission (IOC) of the UNESCO spearheaded ocean-related activities and programmes. Policy infirmities have certainly hindered decisions on issues relating to the Indian Ocean. The contrast is evident when considers the policy attention devoted to the Pacific Ocean. The history of natural disasters shows that the community around the Indian Ocean has suffered consistently. It is a moot question whether Indians feel a sense of complacency about the possible disastrous effects of the ocean. The country however, has responded positively to the creation of coastal zones worldwide, often called Coastal Regulatory Zones (CRZ). The coastal areas of South East Asia ought to be studied more closely. Malaysia has lost 20 per cent of its mangrove forests, Philippines lost 45 per cent and Thailand lost 27 per cent. Even the mangrove forests of Indonesia were adversely affected by the 2004 tsunami. Recent findings indicate that marine life is not as resistant as it used to be before the tsunami.

Over the years there have been a vast number of studies of the Indian Ocean. The ocean is landlocked to the north by Asia and is bounded in the south by the frozen mass of Antarctica. The meteorology of the Indian Ocean is largely conditioned by the solstices of the Tropic of Cancer and the Tropic of Capricorn. The effects of changing seasons over the open ocean and over the interior of a continent need to be considered separately. The Indian Ocean is the only ocean that does not extend from the polar region of one hemisphere to that of the other, being blocked on its northern side by continental Asia. The motion of the sun across the equator influences the temperature of the Asian land mass. The northern Indian Ocean between the equator and Asia stores the heat of the sun within a small distance of the ocean cover. The air over the land becomes warmer than the air over the neighbouring ocean; thus one finds an air pressure gradient between the sea and the land which makes the surface air flow from the sea to the land. This pattern persists during April-September each year. One can attribute the monsoons in summer and winter over Asia to the Indian Ocean. The celebrated British scientist Edmund Halley published a paper in the *Philosophical Transactions of the Royal Society* (1686) and his classic explanation of the mechanics of monsoons is often cited even today.

The World Meteorological Organization (WMO) published an excellent document on the meteorology of the Indian Ocean. Nonetheless, this report does not faithfully or accurately represent the reality of natural disasters. One can ascribe monsoonal wind regimes to heavy downpours in narrow bands extending up to the Gulf of Thailand and Cambodia. One must mention the studies of monsoon rains and the atmospheric condition of the Indian Ocean conducted by the India Meteorological Department and the Batavia observatory about a decade before the tsunami. Winds generate waves and ocean currents and they influence the horizontal and vertical motions of the cold and warm water of the ocean. Being seasonally influenced, the latter assume a monsoonal character. A landmark in investigations of the Indian Ocean was the International Indian Ocean Expeditions (IIOE), sponsored mainly by the IOC during the late 1950s. The IIOE attached great importance to meteorological studies of the Indian Ocean and the concept of general atmospheric circulation in the Indian Ocean and the relationship, or energy exchange, between the ocean and atmosphere were defined during the 1950s.

The WCDR reiterates that poverty reduction should be on the agenda for the future, partly because poverty is closely associated with vulnerability to disasters. Development is often defined rather narrowly as betterment. If the process of amelioration can be prolonged over a period, it is termed 'sustainable development'. Deviations from the process of development usually result in vulnerability. The recent tsunami certainly exposed the coastal vulnerability of Banda Aceh, Bangkok, Sri Lanka and the Andaman and Nicobar Islands. This is distinctly different from floodplain vulnerability or seismic vulnerability. In order to reduce coastal vulnerability, disaster mitigation needs to be looked upon chiefly as a developmental tool. As vulnerable coastal communities' access to food, shelter and opportunities for livelihood are cut off by natural hazards, the links between vulnerability and underdevelopment become evident.

Certain models have been developed for studying natural disasters related to oceans, such as tropical cyclones. These models deal with storm surges like the ones in the coastal districts of Andhra Pradesh in India. The first numerical model to predict storm surges in the Bay of Bengal was developed by P K Das in 1972, and was followed by S K Ghosh's model in 1977. Sections of the Indian coastline may be classified into categories based on the combination of storm surges and wind waves. Stimulated by the disastrous cyclone of 1996 which struck the coast near Kakinada (Andhra Pradesh), recent exercises have tried to estimate the vulnerability of coasts to cyclones and associated storm surges. Being the smallest ocean, the Indian Ocean possesses some remarkably different features. For example, the seasonal reversal of the Somali currents depends on the wind direction of monsoon. Since the northern boundary of the Indian Ocean is at considerably low latitudes, intermediate and deep water can hardly be formed in the northern Indian Ocean. Thus heavy water has to come from the southern ocean, the Mediterranean Sea and the Persian Gulf. As warm surface water flows to the Indian Ocean from the Pacific Ocean, there is an inter-oceanic exchange of sea water. Many scientific questions are raised about the Indian Ocean, particularly about the oceanic processes that generate and maintain Sea Surface Temperature (SST) anomalies.

Sediments, sand and mud are derived from the rivers and the coastal area itself. The beach width and slope vary according to changes in wave energy. High-energy waves and strong winds form a narrow steep beach covered with rough sand and rubble. On the other hand, smaller waves and gentler winds in the summer form a broad gentle sloping beach covered with fine sand. Coastal erosion is a significant natural process and areas experiencing it are easily identifiable. Shoreline erosion results primarily from storms. Coastal erosion claims many acres of beach and wetland in India.

Chapter 5

South East Asian Milieus: Climatic Changes

Climatic change is a very broad concept, including global warming and rises in the sea level among numerous other phenomena. Ideas about global warming and rising sea levels have evolved over decades. Since the basic elements of these two phenomena are the atmosphere and the ocean, one cannot preclude the possibility of the escalation of circulation and air-sea interaction leading to tropical storms, thunderstorms, storm surges, etc. As the El Nino is often responsible for natural disasters, the relation between ENSO and climatic changes needs to be closely studied. Climatic change is related to vulnerability because of physical and social factors. The patterns, severity and frequency of climate-related natural disasters change continuously, and these in turn are likely to endanger the community. It is imperative to keep society immune from the two phases of natural disasters—(1) prior to their occurrence and (2) their aftermath. Of course, restoration and renovation are essential during the post-disaster phase. The impact of climatic change is becoming progressively more financially damaging because of frequent tropical cyclones, loss of land as a result of the rising sea level and damage to fishing stocks, agriculture and water supply. By 2050, some low-lying states like the Maldives may incur losses of almost 10 per cent of their GDP due to climatic change. These startling figures were revealed by the UNEP in order to ensure that higher priority is accorded on the international agenda to natural disasters caused by climatic changes.

Despite various efforts to reduce the emission of Greenhouse Gases (GHG), their reduction hasn't reached the desired levels. Human-induced climatic change may be observed in parts of the developed world. Several developing countries seem quite reconciled to the fact that they have to adapt to the consequences of global warming. An Intergovernmental Panel on Climate Change (IPCC) report, sponsored jointly by the UNEP and the WMO, may be cited to emphasize the urgent need for quick action. The IPCC contends that temperatures across the world could climb by 1.4-5.8ºC in the twenty-first century. The UNEP insists on clean energy technologies so that rising sea levels, changing rain patterns and the impacts of global warming can be better observed. The insurance sector, concerned about the estimated costs of climatic change, is trying to draw governments' attention to the damage caused by increased GHG in the atmosphere. The cost estimates of Munich Re, one of the world's biggest re-insurance companies, are based on the assumption that carbon dioxide concentration would rise to pre-industrial levels in 2050. In 1997, Munich Re estimated that with respect to global climatic change 'the economic cost of weather-related events amounted to US $98 billion with an associated insured loss of US $12 billion'.[1] In an article published in *Our Planet*, a UNEP magazine, the chief of Munich Re's geo-science research group Dr Gerhard Berz said that there was reason to fear that climatic change would lead to natural catastrophes of hitherto unknown force and frequency. According to him, climatic changes could cause worldwide losses of many hundreds of billions of dollars per year. For most countries such losses may vary from a few tenths of a per cent to a few per cent of the GDP each year and the same is true for the small island states. Flood protection schemes to counter rising sea levels and storm surges may cost an average of US $1 billion per year; for low-lying areas like the deltas of Bangladesh and small island states like Maldives, the cost could be significantly higher. Undoubtedly the ecosystem will suffer enormous losses.

According to some costing exercises, natural disasters (including frequent cyclones and hurricanes) could perhaps add a further US $3 billion to costs relating to climatic change. Dr Berz's article provides a break-up of possible costs for different parts of the world, e.g. China's biggest losses appear to be around US $8 billion, mostly in the agricultural sector. According to Dr Berz these can be considered as future risks which need to be provided for through a long-term cover for natural hazards and also through measures for reducing the build-up of GHG. Climate-related natural disasters add a new dimension to the urgent need for an early warning system in the Indian Ocean. Fragile coastal areas where communities are more vulnerable to natural hazards must be identified. Sensitive and disaster-prone areas must be mapped on a global scale. Daniel Claasen, a member of the UNEP team, has developed what he calls a 'vulnerability index' to provide the relevant authorities with a measure of how disaster-prone certain areas are. According to him, a hilly area prone to soil erosion may occupy a high position on the index. A site may be especially vulnerable to mudslides and landslides as a result of heavy downpours caused by El Nino. The destruction of mangroves and reefs in coastal regions exposes inhabitants to greater risks from floods and storm surges. According to the UNEP, any programme to develop an early warning system must use the help of the vulnerability index.

Global warming has affected China's coastal environment in two ways. First, the heightened sea level has inundated vast expanses of coastal areas, causing both salt water intrusion and coastal erosion. Second, typhoon activities have intensified; the latter is often confused with the mobility of typhoon tracks in the north. In a paper ('Impact of Global Warming and China's Coastal Environment') presented at the International Seminar on Climate Change Impact in Beijing in 1991, Prof. Han Mukang discussed some critical problems concerning the rise in the sea level off China. The rise was found to be greater along the low-lying coastal plains in different parts of China, e.g. along the North China Plain near Tianjin and the East China Plain near Shanghai. The localized acceleration in the sea level rise along these plains may be attributed to both tectonic activities and the excessive use of groundwater, as witnessed in the downtown areas of Tianjin and Shanghai over the last few decades. Global warming is a probable cause of the reported northward shift of typhoon landing sites since the early 1970s.

Researchers feel that the global sea level rise projected by the IPCC presents a serious topographical and socio-economic threat to China's coastal areas. The low-lying coastal plains constitute one-fifth of China's coastal area and every year typhoon-induced surges cause heavy flooding and a sea level rise of around 1.5 to 5 metres. Prof. Han Mukang has also indicated the serious consequences of the sea level rise. According to him, four main coastal plains covering a vast area would be inundated. There would be severe salt-water intrusion along the northern China coastline near the Bohay shoreline and the coastal area of Laizhou Bay, and several beaches would be washed away due to coastal erosion. According to Prof. Mukang, 'An appropriate countermeasure to the sea level rise on these plains is to raise and consolidate the existing dykes which have been built in almost continuous span along the whole coast.'

The investigations described above have raised some fundamental questions. The onset of natural disasters may be slow or rapid, but the process of recovery is never quick. Indeed, climatic changes occur over long periods of time and hence they can hardly be described as 'sudden onset, sudden recovery' phenomena. There is a fallacious tendency to use the phrase 'flip-flop' to describe climatic changes. As William Calvin points out, an abrupt climatic change may be caused by a sudden natural phenomenon, but the climate does not flip back to its original state soon afterward. Instead, long-term climatic changes are set into motion.

The eruption of Mt Pinatubo in the Philippines in June 1991 is a case in point. The largest eruption of the twentieth century, Mt Pinatubo ejected more than 30 trillion pounds of rock, dust particles and gas into the air. Although the dense plume of smoke disappeared in a few days and the dust clouds were dispersed and the sky cleared after a few months, Pinatubo had left behind about 40 billion pounds of sulphur dioxide gas which rose to the stratosphere immediately after the eruption and reacted with oxygen and water vapour to form a haze of sulphuric acid droplets. Though invisible to the naked eye, one-third of this mist was still suspended in the air about two to three years after the eruption. Volcanic clouds of sulphuric acid act as a planetary shade, reflecting sunlight back into space and thereby cooling the planet. Although the cooling effect is usually lost in the earth's constant weather fluctuations unless the effect of an eruption is truly titanic, the might of Pinatubo's eruption caused a perceptible change in the weather pattern. Publishing their forecast six months after the eruption, James Hansen and his colleagues at NASA's Goddard Institute for Space Studies in New York claimed that the sulphuric acid cloud was so large that the shadow it cast would do far more than block out any warming caused by humans; it would cause a worldwide temperature drop of about one degree by winter 1992. 1993, they felt, would see gradual warming and by the mid-1990s the man-made greenhouse effect would be restored. Hansen's team was right: the planet did indeed record its coolest temperatures in 1992 and a slow recovery began in 1993. The model also accurately forecast that summers would be noticeably cooler in the continents' interiors, and that was precisely what happened in 1992 and 1993, especially in the American Midwest. While the interiors normally heat up quickly under the summer sun being very far from the moderating influence of the oceans, Hansen predicted that they would cool much faster under the influence of Pinatubo's eruption.[2]

One may consider the salt circulation related to paleoclimatic changes and the consequent threat that the ocean shore might be flooded. Although global warming may be the primary focus of studies of climatic change, all the important factors controlling climate and ocean currents (namely evaporation, the albedo effect, etc.) need to be identified so that their strengths and shortcomings can be estimated through computer modeling or simulation. The stability of climatic change can hardly be ignored if we are to investigate its impact on natural disasters. Any insightful analytical process will reveal certain 'thresholds', determined chiefly by the above factors. Thresholds are not merely boundaries between one mode and another, they are also conceptual spaces that help to identify abortive responses and rapid fluctuations between modes, which commonly appear in non-linear systems. If a bi-tabular system is to be maintained, its contents should strictly deal with one state or another, not with the transitional 'threshold'. In reality, an unusual climatic variation such as El Nino can abruptly push climate into an extreme and arbitrary state. As soon as a climatic shift shows signs of escalating into a natural disaster, the potential disaster mitigation strategies, particularly the carbon reduction strategy, must be assessed. The concept of 'stabilization wedges' is found to be useful. Stabilization means that there is no further build-up of carbon dioxide in the atmosphere because emissions are balanced by land and ocean sinks. Undeniably, the concentration of carbon dioxide in the atmosphere has increased by more than 30 per cent in the last 250 years because of human activities. Hence various strategies of carbon reduction are necessary. If we assume that natural sinks of carbon in the ocean and on land will continue as they have in the past, we ought to achieve a sort of flat trajectory growth for no carbon and low carbon energy strategies on a massive scale, particularly in developing countries.

According to the National Centre for Atmospheric Research, the percentage of earth's land area affected by severe drought more than doubled between 1970 and early 2000. The fraction of arid land (described as -3 or less on the Palmer Drought Severity Index), rose from about 10-15 per cent in the early 1970s to about 30 per cent by 2002; almost half of that change is due to the rise in temperature rather than the decrease in rainfall or snowfall. Global climate models predict widespread drying because of the increase of GHG and carbon dioxide. By factoring out rainfall and snowfall, it may be concluded that the rainfall deficit is chiefly responsible for the increase of dry soils in Asia.

Global climate modeling or the modeling of ocean circulation may be unable to predict the consequences of tampering with down-welling sites. But there is little doubt that better modeling coupled with sophisticated instrument packages sensitive to ocean depths, and satellites measuring variations in the sea surface will make it easier to find a solution. Alternative solutions are being sought, but the human element is now assigned great importance in each case. Any sudden climatic shift is likely to disturb food-supply pathways.

Natural disasters such as tropical cyclones, earthquakes or even tsunamis may appear less formidable than sudden cooling. The latter is different from natural disasters; projections indicate that cooling will affect much of the earth in future decades. The recovery from this phenomenon will take a very long time. Volcanoes spew sulphates and these reflect some sunlight back into space, but we cannot assume that this will counteract the longer-term carbon dioxide emission. In addition to the hitherto predicted consequences of global warming—namely stronger storms, methane release, rising sea levels and stronger El Ninos—we must consider the consequences of abrupt cooling.

No discourse on climatic change and natural disasters can be complete without considering the vulnerability of the affected people and places. Developing countries and particularly the least developed countries are undoubtedly the most vulnerable to climatic change. As mentioned earlier, there exists a 'differential vulnerability' and this is vital for understanding the essential nature of climatic change. The vulnerability of human and ecological systems to the stress exerted by climatic change has not been properly evaluated. Apart from issues of equity and social justice, the reduction of human vulnerability, the enhancement of resources to cope with disasters, the development of the capacity for adaptation and the strengthening of resilience should be considered as a whole. The assessment of vulnerability thus lies at the core of studies of climatic change.

NOTES

1. Salt, Julian E, 'The Insurance Industry: Can it Cope with the Catastrophe?' in Pelling, Mark, ed, *Natural Disasters and Development in a Globalizing World*, Routledge, London, 2003.
2. Zimmer, Carl, 'Pinatubo Lives: Volcano Blamed for Global Temperature Drop, Ozone Depletion', *Discover*, January 1994 [http://www.findarticles.com/p/articles/mi_1511/is_n1_v15/ai_14902729].

Chapter 6

Early Warning Systems

THE MINISTERIAL MEETING AT PHUKET

Although the WCDR and the ASEAN summit in Indonesia built up a consensus for setting up an early warning system in the South East Asian region, the Declaration on Regional Cooperation in Tsunami Early Warning was formalized at the ministerial meetings in Phuket, Thailand on 28-29 January 2005. Cooperation was sought on a regional basis; emphasis was laid on collaboration with global bodies, thereby facilitating international cooperation. The IOC under the UNESCO and the Global Earth Observation System of Systems (GEOSS) are two important organizations that help mitigate the impact of tsunamis and other disasters. The GEOSS is mainly concerned with seismicity and thus it deals with tsunami warning systems. Many countries around the world have seismographic networks through which seismic waves can be monitored and recorded. Some of these networks share data with other countries. Such a network can have a number of stations in different parts of the globe, capable of recording nearly all the earth's seismic signals. It can thus be used for tsunami warnings and other related purposes. Global Seismographic Network (GSN) data belonging to the US is easily accessible to any user. The International Federation of Digital Seismic Networks and the International Monitoring System (IMS) have similar operations. The IMS is much more than a seismic network and has a number of stations to monitor underground phenomena. IMS' seismic monitoring system has 50 primary stations that provide data to the International Data Centre (IDC) and also 120 auxiliary stations to allow data access on request from the IDC. Although the IMS is not formally designated as an early warning system, it could record a northern Sumatra earthquake on 78 of its monitoring stations within a minute of the event. Consequently, within two hours of the earthquake the IDC could notify countries about the disaster.

Broadly speaking however, the IDC is insufficiently equipped to issue earthquake warnings, mainly because its processing system is incapable of detecting potentially destructive earthquakes or tsunamis. It is felt that IMS data should be used for tsunami warnings by sending it to international organizations such as the UNESCO's IOC. At the ministerial meeting at Phuket, ministers and special envoys from 42 countries and the European Commission formulated a declaration whose 18 points developed a framework and timeline for an early tsunami warning system. The Declaration on Regional Cooperation on Tsunami Early Warning Arrangements calls for an early warning system that should necessarily 'include risk assessment, hazard monitoring and detection, prediction and formulation of warning, discrimination and communication of warning messages and knowledge and preparedness to act'. The Declaration set the target date for implementing the essential elements of a regional tsunami early warning system and stressed the involvement of global institutions. The meeting foresaw the application of a higher level of technical sophistication that would allow several regional tsunami warning systems to share data about the Indian Ocean region. There is an International Tsunami Warning System for the Pacific. The physical elements of operational early warning systems for tsunamis in the other parts of the world consist of sensors in the sea, on land and underground (to record land motions), transmitters linked to satellites, the Internet and devices to relay the changes recorded to other stations around the world. The use of advanced technology is a key factor in designing early warning systems that can help protect lives and property from the devastation caused by tsunamis. An important element is the Deep Ocean Assessment and Reporting of Tsunamis (DART) which currently comprises six moored buoys providing timely and accurate tsunami warning information. There is also the system of tide gauges that provides data essential for early warning systems. There are instruments to monitor land masses, i.e. to measure and record the vibrations of the earth through seismic stations; this is precisely what a GSN does. The term 'subduction' refers to the process by which an ocean plate slides under a continental plate, usually at the rate of a few km per year. Naturally, the friction between such plates could cause tremors, but the plates may lock against each other leading to a compression of the sea floor, causing an overlying coastline to strain upward. As the released pressure finally pops, the sea floor lunges landward and the coast plunges seaward. Naturally, property and real estate by the seaside are the first to be destroyed. The shifting plates cause the seawater to spread violently in all directions, often resulting in a tsunami. In Indonesia seismic activity occurred in a subduction zone, measuring 9 on the Richter scale. Indeed the rupture of the South Asian subduction zone on 26 December 2004 was the first major test of GSN technical design goals which scientists had formulated over 20 years ago. Seismic waves (essentially waves of energy caused by a sudden motion of rock) from the Sumatra earthquake reached a monitoring station in Sri Lanka within four minutes. Within just 21 minutes the needles on thousands of seismometers worldwide registered violent motion. Scientists feel that the ultimate goal should be to improve 'global coverage' by collaborating with the oceanographic community to have stations on the sea floor. Although this is technically difficult to achieve, stations on the sea floor will certainly be a major step forward. This sums up in brief, the salient points of the Phuket meeting.

DEVELOPMENT OF EARLY WARNING SYSTEMS

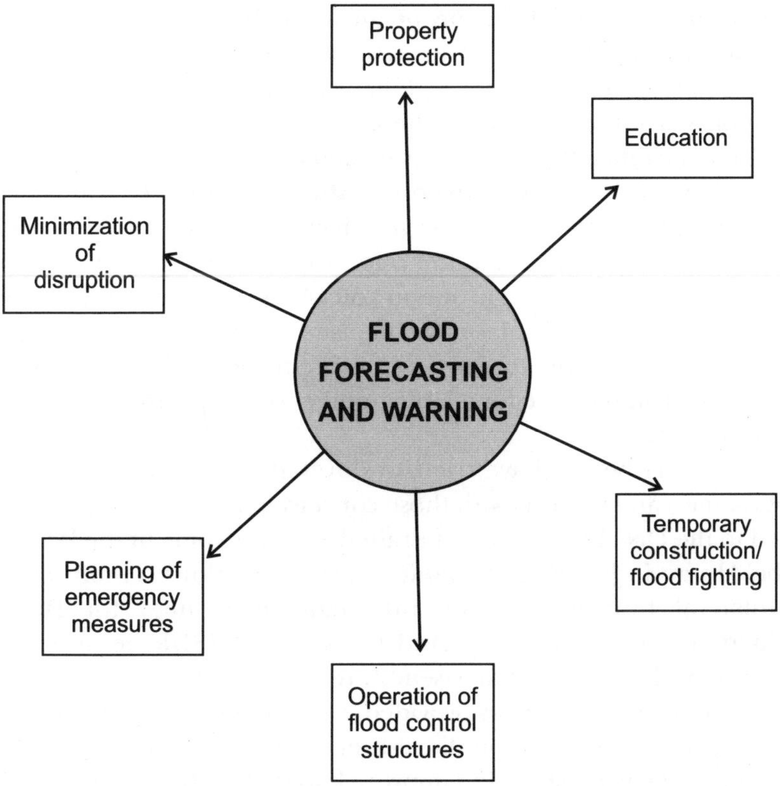

Fig.17 Benefits of flood forecasting and warning

The 2004 tsunami that claimed more than 150,000 lives has exposed the vulnerability of the Indian Ocean's coastlines. All subsequent international meets have accorded the highest priority to early warning systems for natural disasters. The matter had been considered and discussed prior to the WCDR in Kobe, Japan. Indeed an International Early Warning Programme (IEWP) had been proposed at the second international conference on early warning in 2003 at Bonn, Germany. The IEWP works to increase resistance to all types of natural hazards and embraces various UN organizations such as the WMO, UNESCO, UNEP, ISDR, etc. Efforts to put an early warning system in place began with the Report on Early Warning Capacities with Regard to Natural Disasters.[1] The report emphasized the importance accorded to timely, well-coordinated and effective early warning systems at the 50th session of the UN General Assembly (December 1995). Thus the UN resolution was not a mere follow-up to the WCDR deliberations. As the scientific and technical committee of the IDNDR pointed out, the efficacy of an early warning system does not lie merely in the detection or forecasting of a possible hazard and the issue of an alert. It was therefore hardly surprising that the UN issued a press release on 19 January 2005, making public its integrated plans for a global early warning system. The basic intention was not only to reduce the death toll caused by natural hazards but also to facilitate the rapid transmission of data, sensitizing people to the risks to which they were exposed. After the Indian Ocean tsunami it was strongly felt that fewer lives may have been lost had there been a tsunami warning centre like the one in the Pacific Rim. The Indian Ocean, like the Pacific, is prone to volcanoes and earthquakes. Tidal gauges, fast transfer and alarm mechanisms and a deeper knowledge of the dangers in sensitive zones may have reduced fatality. These would have allowed ample time for a large number of people to seek refuge in higher places.

Several UN bodies have tried to dovetail the IDNDR and WCDR programmes and projects with those conceived by various organizations allied to the UN. The document drafted as an outcome of the WCDR at Kobe, Hyogo in Japan (18-22 January 2005) mentions certain relevant multilateral frameworks. The International Strategy for Disaster Reduction (ISDR) often referred to as the UNISDR because of its adoption by the UN General Assembly resolutions 59/L10, 58/214, 57/256, 56/195 and 54/219 has provided 'the United Nations' system with the mandate to promote public awareness and commitment, expand networks and partnerships and improve knowledge about disaster causes and operations for risk reduction'.[2] Other UN resolutions encouraged the Intergovernmental Panel on Climatic Change 'to continue to assess the adverse effects of climatic change on the socio-economic and natural disaster reduction systems of developing countries'.[3] The Plan of Implementation formulated at the world summit held at Johannesburg in 2002 had also drawn attention to the need for '[. . .] an inclusive, multi-hazard, inclusive approach to address vulnerability, risk assessment and disaster management, including prevention, mitigation, preparedness, response and recovery, as an essential element of a safer world in the twenty-first century, supporting the International Strategy for Disaster Reduction as the first action.'[4] The Millennium Declaration identified the key objectives of 'protecting the vulnerable', 'protect[ing] our common environment' and 'intensify[ing] cooperation to reduce the number and effects of natural and man-made disasters'.[5] This Declaration is deliberated upon by UN spokesmen even today.

As the early warning system was essentially concerned with the Indian Ocean, the UNESCO's IOC in Paris in its 23rd Assembly from 21-30 June 2005 formally launched the Indian Ocean Tsunami Warning System (IOTWS) by adopting a resolution to create an Intergovernmental Coordination Group (ICG/IOTWS) to play a leading role in developing and overseeing the early warning system. This group will organize data exchange, promote tsunami research, establish and develop national tsunami warning and mitigation capacities and implement a multi-hazard approach to the warning system. Preparing a framework for an early warning system for ocean-related hazards has thus acquired global importance.

The Yokohama Strategy concluded that early warnings were a crucial element of disaster risk reduction. There is a greater awareness of the importance of early warning systems than ever before. With advances in technology, the technological capabilities of early warning systems need to be greatly improved—computer-based prediction has rapidly improved because of advances in communication technology. However, there are still some countries whose policies on early warning systems have not kept pace with technological advances and thus—as demonstrated by contemporary events—millions of people continue to be at risk. Nevertheless, it is now universally recognized that risks, vulnerability and sustainable development are all inextricably interwoven. At its Bonn conference the IEWP emphasized that early warning systems must feature prominently in national plans and policies. The WCDR strongly recommended implementing the directives of the Bonn conference and the Mauritius Strategy for developing countries.

It is necessary to explore the mechanisms that a tsunami early warning system should be equipped with to avert the destruction of lives and property. Despite the bulk of data about the Sumatra earthquake, timely safety measures could not be taken. The crucial point about the warning process is the detection of the tsunami immediately after its genesis in an earthquake. There would have been far fewer casualties if the event had been purely seismic. Detecting and monitoring the course of a tsunami at its inception is imperative for early warning. Sometimes however, it may not be possible to gauge or correctly predict tsunamis. At the Pacific Tsunami Warning Centre (PTWC) at Hawaii, for example, early alert signals were false on certain occasions.

The Tsunami Newsletter continues to alert civil society about threats posed by tsunamis. The assessment of these threats requires rigorous computation using high-quality seismic data before the warning centre issues an alert. There are a variety of parameters which need to be estimated but there is likely to be an error gauging the tsunami's height if there is an error elsewhere in the available seismic data. Computer models are often developed by calculating the propagation of tsunami waves. A tsunami may pass absolutely unnoticed since its slope is greater even though a single wave may be as long as 750 km. What then is the first indication that an earthquake may have caused a tsunami? In an article on tsunamis vis-à-vis earthquakes in the Tsunami Newsletter, Dr Charles McCreery says that one has to wait till a potential tsunami reaches the nearest sea level gauge to confirm or to deny its existence. Since the Pacific Centre has 100 such gauges, it is possible to transmit the data via satellite back to the warning centres. These gauges are located in harbours and protected bays. As the characteristics of tsunamis change because of shallow depths, the data from the gauges does not always indicate the tsunami's arrival. Therefore, one uses the DART gauge, which has a highly sensitive pressure recorder installed in the ocean. The recorder can detect the altered pressure even if the height of the ocean above it changes by one centimetre. The reading is transmitted acoustically to a surface buoy that in turn relays it via satellite to the warning centre. DART gauges are situated in deep water so that they can accurately record tsunami waves as they pass. Besides DART buoys in the ocean, other equipment such as tide gauges are used. These are devices to measure the tides and if equipped with a radio transmitter, data can be transmitted in real time. A tide gauge may be described as a small tube containing sensors that measure water levels; the National Oceanic and Atmospheric Administration (NOAA) tsunami warning centres use about 100 water level tide gauges, which are often installed on various structures in the ocean such as piers, lighthouses and oil drilling platforms. A large tsunami wave in motion can easily fill the tube; thus some of them are built to withstand subsequent waves. In fact nearly 100 are found around the Pacific; 33 of them transmit data within five minutes of receipt and the water level from these 33 gauges can be sent directly to centres for seeking further information. There are many coastal gauges in other countries around the Pacific. A tsunami slows down but increases in height as it approaches the shore. It strikes the shore as a breaking wave which resembles a wall of water or a tidal flood. One can cite a host of theoretical studies which provide excellent insight into the mechanics of tsunamis. Such pursuits are often termed 'modelling'. Tsunami waves two to three metres high can cause widespread damage. A critical function of modelling is to ascertain the height of a tsunami in the last stage of its evolution. The NOAA has developed the computer model MOST (Method of Splitting Tsunami) which essentially simulates the generation of a tsunami, its trans-oceanic propagation and inundation of dry land. Indeed, a knowledgeable tsunami modeller is needed to take care of effective tsunami warning and forecasting capabilities. Supercomputers have been used to develop virtual tsunamis. Unfortunately some countries which receive tsunami warnings from international agencies find it difficult to act upon them because of the lack of emergency response mechanisms and appropriate communication infrastructure. Despite receiving a warning, coastal communities are sometimes unable to escape the wrath of the tsunami. Satellite-based cyclone warning systems are here to stay; perhaps they may be augmented by overhauling the existing infrastructure that uses DART gauges.

On the whole any hazard warning system must consist of four essential components: (a) the detection and observation of the hazard (b) the preparation of the message for forecast and warning (c) dissemination of the warning (d) preparedness and response. These components are interdependent; of course, component (c), i.e. the dissemination of the warning is of paramount importance. To be truly effective a warning system must incorporate state-of-the-art communication techniques. Warning systems for all types of natural hazards ought to be in place. Unfortunately most systems have been developed ad hoc and their development seems contingent on the warnings of meteorological organizations, media broadcasts or government disaster response agencies. The maintenance of a warning system should be entrusted to an apex body, without underplaying the creative thought that often determines the precise nature of the links in the chain of warning. One must clearly identify the function of each component rather than focusing excessively on the means of communication. The process is often retarded by confusion amongst the so-called originators, intermediaries, disseminators and the recipients. The problem is compounded if there is a non-identifiable head of organizations with the sole responsibility of ensuring that the warning is rounded off with protective action to save lives and property. The long coastline of South Asia demands an enormous capacity for coordination. It is impractical to entrust a network to a single authority, particularly with respect to the communication of natural disasters. Even though we may not have comprehensive solutions to such problems, we must have an overall authority (a single person or a closely integrated team) to maintain controls and standards. Supporting laws ought to lay down minimum standards for communication and dissemination. Data from past and present events ought to be fed continuously into the system, thus constantly updating it. Virtual tsunamis need to be realistically simulated.

One may refer to a recent global deliberation by the Tropical Cyclone Panel of the Geneva-based WMO. There is a list of the names of cyclones in parts of the world susceptible to storms. While hurricanes in the Atlantic and Pacific area are listed alphabetically, the cyclones in the north Indian Ocean zone have not been mentioned. Even though some cyclones in China, Japan, Bangladesh, Sri Lanka and India originate in the Indian Ocean, it is curious that their names do not feature on the list. It is now almost two years since the tsunami and there are bilateral efforts to monitor seismicity and tsunamicity in the Indian Ocean. The India-Indonesia joint venture is an important step and has significant implications for the future. One hopes that this will be more than a routine exercise on surveillance and patrolling.

The tsunami early warning system envisaged for the Indian Ocean has raised many questions about existing systems. Critical questions regarding an infrastructure for the timely reception of tsunami warnings have been raised because recent emergency communications have proved virtually useless in many instances. Policymakers ought to be well briefed in order to understand the implications of the Indian Ocean tsunami and the current capacity for detection and early warnings. Although it poses serious challenges, the establishment of an extensive tsunami early warning network in the Indian Ocean is expected. It is felt that a global tsunami detection/warning system ought to be set up in the Indian Ocean, consolidating similar systems in the Pacific and identifying/developing possible sites in the Indian and Atlantic Oceans. The costs (including those of infrastructure and emergency management) need to be assessed. A DART buoy system is essential; it must have features that allow it to serve as a critical warning system for the Pacific, Indian and Atlantic Oceans. The NOAA's DART programme serves the different phases of early warning systems. The components of the DART system are shown in Fig. 18.

The NOAA'S role in the World Weather Watch (WWW) of the UN is well known; the WMO also administers cooperative programmes that deal with scientific research, weather data collection and management, meteorological forecasts and warnings. WWW operations are sustained

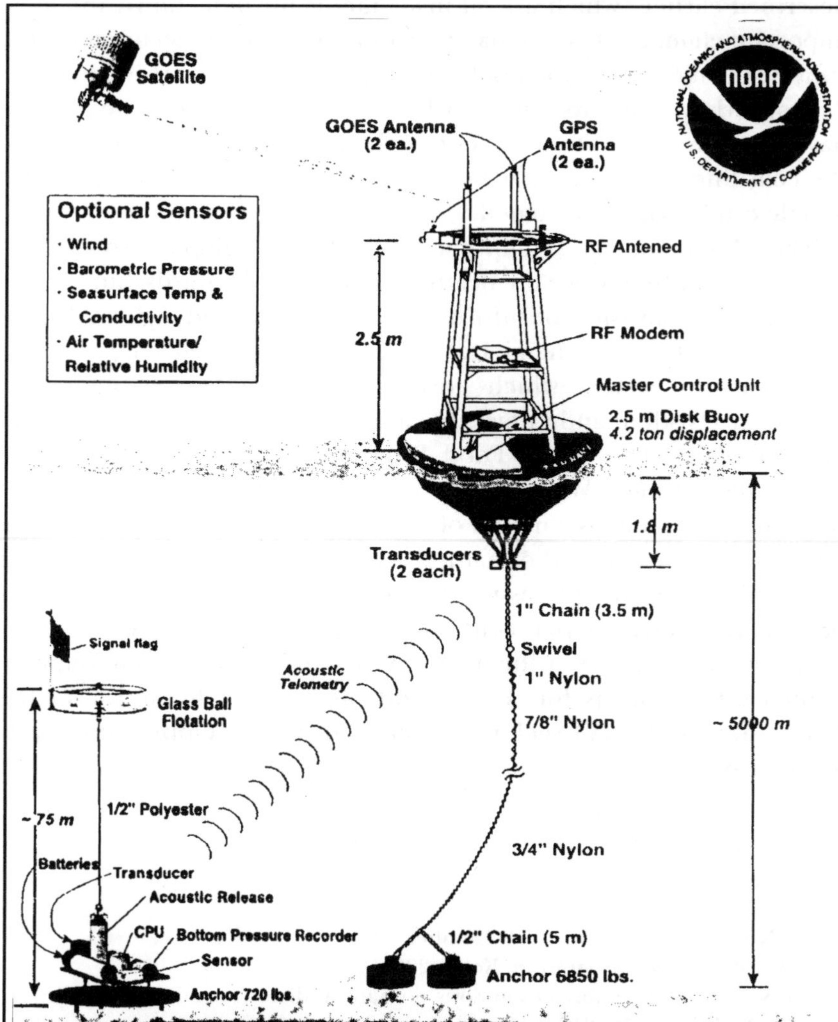

Fig.18 NOAA DART Platform, Tsunameter Mooring System

globally through a well-established international telecommunications network for receiving and distributing weather data and warnings. Two WWW data centres are used for weather data analysis and forecasting. Indeed, collaborations are essential for building effective new mechanisms.

For a tsunami warning system, advanced technology is undoubtedly important. The first step should be to have seismic networks in place to pick up the first signal of an earthquake that might generate a tsunami. Apart from the tide gauges, satellite telemetry and ocean buoys described earlier (which are all instruments for detection), the most important element of any tsunami warning centre is the technology that conveys the message to the public. Post-tsunami activities should tap the data of early warning systems in order to understand where the tsunami has struck, measure the water depth (bathymetry) and topography of the coastline and assess how far inland flooding might occur so that people can be safely evacuated.

The UN has launched comprehensive plans for a global early warning system not only to reduce the deaths due to natural hazards, but also for the rapid transmission of information to regions endangered by the recent Indian Ocean tsunami. It is now clear that we need a multi-hazard early warning system which ensures environmental stability, and indigenous wisdom and knowledge also need to be incorporated into disaster plans. The World Food Programme (WFP) generated the Humanitarian Early Warning Service (HEWS) web to bring under a single umbrella the vast amount of information available from technical institutions on each kind of natural hazard. One-third of more than a hundred million people assisted by the WFP have been affected by natural disasters, 90 per cent of which are of meteorological or hydrological origin. Over the next 15 years, the WMO seeks to improve upon 'alerting systems' for weather- and water-related disasters. Even the UN Volunteers (UNV) seek involvement in the development of early warning systems.

NOTES

1. UN Document A/50/526, 9 October 1995.
2. WCDR, Hyogo Declaration, Kobe, 2005.
3. UN General Assembly Resolutions on natural disasters and vulnerability, A/59/XXLII and 58/215.
4. WCDR Conference, 199/20.
5. UN General Assembly Resolution 55/2, September 2000.

Chapter 7

Impact of Hazards, Vulnerability and Development: A Critique

INTRODUCTORY REMARKS

Both the IDNDR and the WCDR have provided a platform to deliberate on actionable steps after the tsunami. Conceptualization and realization are symbiotic and disaster mitigation is now looked upon as a development tool. The Hyogo Declaration clearly indicates that the reduction of vulnerability and the impact of hazards are prerequisites for development. While the WCDR attached importance to development, the need for sustainable development did not feature prominently in the declaration. This chapter outlines alternative means of development necessary when considering sustainability. The ways in which the impact of natural hazards may be reduced has been discussed, followed by an examination of development vis-à-vis vulnerability. Is it possible to formulate developmental alternatives in the first decade of the new millennium?

REDUCING THE IMPACT OF NATURAL HAZARDS

Developing countries take much longer to recover from natural disasters than developed countries. Of course, the process of recovery in both developed and developing countries shares many common features. In a matter of minutes, the 2004 tsunami reshaped and redefined the shorelines of the Indian Ocean. On an average, Bangladesh loses more than 100,000 lives to natural disasters every year. Typhoons similar to Typhoon Talim in China and Typhoon Nabi in Japan (both occurring in 2005) are very frequent; the death toll has been greatly reduced because of phenomenal advances in technology. But the figures are still frightening, running into millions.

Scientific and technological practices vary from region to region and nation to nation. In reality there is still a disturbing gulf between risk and preparation. Disaster warnings and forecasts are still assigned secondary importance in some disaster-prone countries. Given the need for instant responses during natural calamities, the local populace ought to be alerted well in advance. It is encouraging that many establishments, countries and states are trying to promote self-reliance, which is really the bedrock of effective responses to disasters. While doing so, it is necessary to identify the collective strengths and weaknesses of the community. The recent tsunami indicated the need to adopt more reliable preparedness and response systems without ignoring social, cultural and geographical diversity, and the latest Government of India status report on disaster management corroborates this point of view. Although most countries in the South East Asian region possess some infrastructure for disaster management, one can hardly claim that their capacities and efficiency levels are uniform.

It is a pity that because of the low priority accorded to disaster management the relevant committees of South East Asian nations met only during emergencies and confined their activities to relief operations. The bulk of relief work is undertaken by NGOs, besides well-known international bodies. The image of the establishments' response systems has not been particularly impressive. Therefore, the design and development of a preparedness and response system is imperative for expanding the scope of relief and disaster management. Any strategy for natural disaster management should strengthen the links between the government and NGOs and improve the understanding of operational strategies. This will go a long way in developing an effective warning activation and community alerting system. It is now clear that the earlier strategy was based on a number of wrong assumptions and impressions. There ought to be a structured set of roles and responsibilities to protect people and their property from impending disasters. Response mechanisms must be developed by focusing on the essential characteristics of hazards and the efficacy of such systems ought to be tested in the immediate aftermath of the disaster. While the evaluation of response mechanisms is a long and slow process, lately some countries have framed operational support plans based on the evaluation process. In certain cases swift responses have provided excellent results. Sophisticated new tools and methods have emerged that require specialized training. An effective response to natural disasters is possible only when it involves the active participation of individuals.

The broad structure of a community consists of individuals, families and social groups that are linked to each other by various intangible factors such as values, relationships, attitudes, etc. Common concerns and interests, when taken in their entirety, can be conceived of as a system or a network. As soon as the community receives the disaster warning it is galvanized into activity because of its inherent links. The next concern is about the impact of the disaster, which the community should work to reduce. After the onslaught of a natural disaster the community actively tries to ensure collective survival. Social values assimilated by the members of the community create a sense of camaraderie and the community—even those unaffected by the disaster—begins to form new networks to pool resources and ensure mitigation of the impact of the disaster. The leadership assumes the role of establishing an effective response recovery system. Once the pre-disaster community structure re-organizes itself, a new ethos emerges on the basis of needs created by the disaster. But the process of recovery, however effective, cannot restore an affected community to its earlier state.

The international community felt strongly about the overwhelming devastation caused by the recent tsunami. The seismic activity in the Indian Ocean set into motion a series of disturbances in different parts of the Indian oceanic system. Almost two years after the event, the process of recovery still continues. The repercussions of the disaster have been so complex that even today communities find themselves unable to function as they did prior to the disaster.

Studies of natural disasters ultimately investigate the possible means of their reduction. Many have dwelt on strategic goals, frameworks and priorities for reducing the impact of disasters. Disaster mitigation entails a generic set of actionable priorities, irrespective of the type of natural hazard. The set comprises (a) the observation, characterization and prediction of natural hazards; (b) the assessment of actual and acceptable risks; (c) the development and implementation of alternative measures for risk reduction; (d) familiarity with the database of earlier natural disasters. Though these priorities apply to all natural disasters, there may be situations where the present state of technological capability and knowledge fails to provide a better understanding of the disaster; hence, the need for research. For example, though the recent tsunami was generated by submarine fault movement it could have been because of submarine volcanic explosions and submarine landslides. The Pacific Rim has recorded several false tsunami alarms. Of course, these must be minimized. Extensive research on the Indian Ocean is necessary and this must include investigations of the relation between the rate and magnitude of the stimuli of disasters, the character of the resulting tsunami and the factors influencing the path and run-up of the tsunami. The response immediately after impact should be tactical and logistical. As natural disasters are often beyond local response capacity, there ought to be professional and volunteer groups trained in on-site emergency procedures.

VULNERABILITY

Vulnerability increases if there is no recovery strategy for a specific hazard. Andrew Maskrey, in his classic work *Disaster Mitigation: A Community-Based Approach* (Oxfam, 1989), has discussed vulnerability from a different standpoint. If we look upon the IDNDR and WCDR as culminations of the global concern about natural disasters, vulnerability too ought to feature high on the agenda. It is now widely held that disaster vulnerability is inextricably linked to the processes that reflect dependence and underdevelopment. Vulnerability to disasters has often been identified with and influenced by, if not related to, other forms of vulnerability. During the 2004 tsunami the relation of ethnicity to disaster vulnerability was evident in the Andaman and Nicobar Islands. The complexities of ethnic heterogeneity were readily discernible. For example, a particular group of the indigenous population anticipated the onslaught of the tsunami and took refuge in safer regions, while another group resisted the rescue efforts and even confronted the rescuers; yet others hid themselves, deliberately creating the false impression that they had been washed away.

The Bhuj and Latur earthquakes testify to the fact that men and women respond differently to both warnings and subsequent phases of disasters. Women seem to have a different perception of risk as well. The growing vulnerability of women to natural disasters exposes pertinent gender issues. Women are found to possess an instinctive capacity greater than men's to recover from the shock of natural disasters. But social disparities often lead to greater suffering for women trying to come to terms with a disaster.

Two days after the tsunami, the extent of death and despair on the shores of the Indian Ocean became known to the international community. Of the 140 deaths reported in Kollam, Kerala, 139 were from the Allappad panchayat alone. As the 10 metre high wave hit the fisherfolk, Rajeevi lost her pregnant daughter, Pinchu, and found that their home and those of their relatives had been destroyed. A 51-year old woman clung to the floating trunk of an uprooted coconut tree to keep herself afloat. After the sea had calmed many fishermen returned to find their wives and children dead. The scenario in Sri Lanka was more terrifying. Southern Galle, a popular tourist spot, was among the first places to be submerged. Soon after, the island of Phuket in Thailand was submerged too. A man on the Patong beach saw the water of the ocean suddenly begin to recede, and in 15 seconds the water had receded as far as 800 metres. His wife had left for a sailboat ride with two other families a while earlier. As the water receded further and further away, there was a momentary calm and schools of fish lay writhing and convulsing on the widened expanse of beach. Several people on the beach saw a gigantic wave in the distance, rapidly advancing towards the shore. A gentleman, his wife and his niece rushed towards a hotel near the beach, while the wave pursued them relentlessly. His wife got to the first floor and he managed to reach the second floor. But their niece had disappeared. The walls of the hotel were battered by boats and cars carried by the wave. Later the frantic couple was told to visit the hospitals to try and find their niece. If she was still missing they could try the morgue. Detailed analysis of the vulnerability of communities is yet to be undertaken. In a remarkable critical investigation, Ian Davis has identified this as a 'key element in a holistic appraisal of who and what are exposed in the threat of a disaster'.[1]

The tsunami plunged the fishing community along the Tamil Nadu coast into a state of panic, but nonetheless the community responded quickly to the alert. About 100,000 members of the community living within 200 metres of the shore of the Nagapattinam district, could be evacuated. The Cuddalore fishing community used to catch about 75 tonnes of fish everyday, but overnight around 29,000 fishermen became virtually jobless in two districts. The restoration of their livelihoods was certainly an immediate challenge. Around 25,000 families of fishermen were endangered. In Cuddalore, 21 children were orphaned and 137 lost one of their parents. At Kottilpad village in Colachel, fisherwomen sought help and support from leading personalities of the area belonging to NGOs like Don Bosco. More than 150 people, nearly half of them children, lost their lives in Kottilpad. Although in Nagapattinam more than 1,000 boats were completely destroyed, 800 partially damaged and in Cuddalore 6,000 boats were totally wrecked, the fishermen and their families hoped to resume their trade somehow. Apparently there was some conflict between the fishermen and the local leadership. The episode and its consequences illustrated the differential vulnerability of the different social strata.

While far greater initiative was required to reduce the vulnerability of the community, unfortunately official agencies did not seem to treat the affected community members or try to identify how support might be provided with any real urgency. The fishermen demanded payments from the government to purchase new boats and equipment and reportedly refused to set out to sea unless the government indemnified losses. The production of new boats and the repair of damaged ones posed problems. Although the government agreed to pay a certain amount, the local community leaders prevented the funds from reaching the victims. The local leaders opposed the government aid as they felt it diminished the community's own sense of self-sufficiency and autonomy. In most households sources of livelihood were in a shambles. The government devised several strategies to alleviate unemployment and poverty. It manufactured mechanized and fibre-reinforced plastic (FRP) boats. Each FRP boat could help provide the livelihood of four or five fishermen while more than ten fishermen could sail in a mechanized boat.

Vulnerability can be reduced if a community-based approach is adopted, intimately drawing upon local knowledge, experience and capacity. 'Vulnerability' is an event-oriented term, but each event that determines vulnerability is unique, changing from occasion to occasion and place to place. The contextual peculiarities of natural disasters are their most distinct feature. The early 1970s witnessed an upsurge of concern about the human vulnerability to natural disasters, but it became an especially significant issue on the eve of the IDNDR. Anderson and Woodrow's landmark book *Rising from the Ashes* (1989) was really a product of the Harvard research project called Capacity and Vulnerability Analysis. According to them vulnerability can be regarded as the principal factor that makes communities more susceptible to calamities. Evidently we need—in the language of mapping—to conceptualize an appropriate range and domain. The combination of 'capacity' and 'vulnerability' could form a range. There ought to be a system for helping to increase the resistance of the population against natural disasters. According to Anderson and Woodrow, an analytical framework provides the mapping which identifies categories of factors and suggests the sequence or order in which they ought to be considered. The relevance of this work becomes evident in the context of the recent disaster in the Indian Ocean. The vulnerability of the Indian Ocean coastline is now well known. Coastal development is a complex issue. From an environmental point of view, coastal vulnerability is a matter of grave concern; thus climatic changes need to be carefully examined. In 1990, Mary Anderson expounded four reasons why disaster-related costs are higher in developing countries.[2]

DEVELOPMENT

Both the IDNDR and WCDR have focused a great deal on development. The processes of relief and development are interdependent and operate in a common environment. Development efforts cover a much broader area than time-bound relief operations. Development responses follow a sequence of actions that reflect organizational layering, or the organization/establishment's hierarchical arrangement. The alleviation of poverty is integral to the process of development. Poverty ought to be understood in a broader sense as not just the lack of money but also the paucity of sources of information. Capacity-building has already been mentioned in this regard. Similarly the risks stemming from the fragility of the local economy need to be identified. The response to natural disasters and the lack of development are often related. When disasters occur in underdeveloped areas they usually destroy the few gains that development had brought. The overriding question remains: shouldn't a region known to be prone to risks be developed?

'Eco-development' is essentially a fusion of developmental processes, keeping in mind the possible degradation of the environment. All developmental processes that aim at disaster preparedness are complementary. Disaster management, if it is not to remain confined to relief and partial recovery, has to be an ongoing process. Although the post-disaster phase begins with emergency relief, it soon moves into the phase of the three Rs—rehabilitation, recovery and renovation—and subsequently into the phases of mitigation and preparedness. All these measures are accompanied by the hope that the next disaster's impact might be minimized. Thus continues the disaster cycle.

The disaster 'continuum' is also often viewed as a military logistics exercise, as some technical skills need to be acquired for developing the interfaces between the phases of the 'continuum'. The perception of disaster cycles as a continuum enables us to see the interconnections with other systems. This approach seeks a sustainable solution to natural disaster reduction. Its multi-sectoral and interdisciplinary nature allows the combined flow of resources and imperatives for development. The word 'triage' may be defined as the principle or practice of allocating limited resources, such as food or foreign aid, on the basis of expediency rather than according to moral principles or the needs of the recipients. Ian Christoplos uses the idea of 'triage' to explain the relation between development policy and humanitarian values.[3] 'Triage' helps understand and makes transparent the choices that risk management must reckon with so that it is possible to identify where and why political will and policy failed in their efforts to mitigate the risk of natural disasters.

The 1970s witnessed an upsurge of interest in the environment and the notion of sustainable development came into being. Despite interpretive divergences, its definitions, goals, indications, values and practices share much common ground. Essentially, sustainable development fuses a concern for the environment with the need for development. The 2002 World Summit on Sustainable Development deliberated on some of the founding principles of sustainable development.

Although the WCDR touches upon this genre of development rather tangentially, it focuses on poverty alleviation. Equity and human development do not seem to have been appropriately emphasized. The IDNDR had a time horizon of a decade while the WCDR has a time horizon till 2015. It is perhaps sheer coincidence that according to the UN Resolution 55/2 (18 September 2002), 2015 is the short-term time horizon of the UN Millennium Declaration. The UN Resolution lays down two sets of goals: the Two Generation Goals (2050) of Sustainability Transition of the Board on Sustainable Development and the long-term (beyond 2050) goals of the Great Transition of Global Scenarios Group. Certain new terms have been coined such as 'sustainability transition' and 'great transition'. The former refers to the provision of materials to feed, nurture, house, educate and employ many more people by 2050, reducing poverty and preserving the basic life support system. 'Great transition' envisages 'a rich quality of life, strong humanity and a resonant connection to nature'.[4] Sustainable development is not only to be defined but also to be measured.[5]

Although disaster ethics have not been granted their due importance by disaster management, the work of humanitarian assistance networks makes the notion of equity indispensable. Sustainability requires the ethical standards or values set down in the Millennium Declaration. As we move towards the time horizon of the WCDR, these values keep assuming newer dimensions. This reflects the culture of shared responsibility which is the quintessence of both the Millennium

Declaration and the Hyogo Declaration. The basic competence of workers and activists in different parts of the world can grow only if they imbibe the values, aims and goals of both declarations.

NOTES

1. Davis, Ian, Proceedings of a Workshop at the Royal Society, London, 19 April 1993.
2. Anderson, Mary B, 'Analyzing the Costs and benefits of Natural Disaster Responses in the Context of Development', Environment Working Paper No. 29, May 1990.
3. Christoplos, Ian, 'Actors in Risk' in Pelling, Mark, ed, Natural Disasters and Development in a Globalizing World, Routledge, London, 2003, pp.95-109.
4. Raskin, Paul, et al, Great Transition: The Promise and Lure of the Times Ahead, report of the Global Scenario Group, SEI PoleStar Series Report No. 10, Boston, Stockholm Environment Institute, 7 March 2002.
5. Parris, T M, and Katar, R W, 'Characterizing and Measuring Sustainable Development' in Annual Reviews of Environment and Resources, No.28, 2003, pp.559-86.

Chapter 8

Desideratum: The South East Asian Context

It is worth referring to the *Natural Hazard Observer*, Vol. XVII, No. 4 (1992) that dealt with the Bangladesh Flood Action Plan (FAP). The FAP was organized by the WB as a consortium of several organizations, following the heavy floods of 1987-8. It had a broad scope and a time frame of over a decade. Yet, it could hardly avoid persistent questions, during different phases, about the role and significance of its components. Did the proposed system of embankments threaten the long-term ecological and natural resource productivity of the delta? Would the embankments increase flood hazards in some areas and endanger some social groups? Had some feasible non-structural approaches been conceived and integrated into the plan? Would social cost and benefit be equally distributed? Did the plan expand the participation and range of choice for those at risk? One might further ask if the FAP represents a rather one-sided approach whose social, political and cultural consequences may be felt later.

Natural disasters cast a ubiquitous shadow on human affairs and they have produced a wealth of chronicles, anecdotes and critical studies. Sometimes what is cataclysmic for one particular group may be considered trivial or unworthy of consideration by another, or sometimes as a significant basis of preparation for future events. Perhaps it is the differential character of the perception of natural disasters that makes them so dangerous. The IDNDR has rendered invaluable service in alerting us to the constituents of a natural disaster, however homogenous or heterogeneous. Although we have a broad knowledge of the characteristics of natural disasters, we can hardly afford to overlook less quantifiable elements of natural disasters such as stress, trauma, anxiety, worry, panic, etc. One ought to refer to the classic work by Prof. Gilbert White on the human adjustment to floods. The 2004 tsunami makes us wonder how people perceive risks in their daily lives and to what extent these perceptions shape attitudes and decisions. It is interesting to analyse why people continue to live in disaster-prone areas even after experiencing the horrors of the tsunami. Are people fully aware of the vulnerability of their locations and how do they equip themselves? On what factors do individuals base their decisions during crises? Undeniably there have been remarkable changes in human adjustment and adaptation, but certain desideratum demand continuous responses about various kinds of disasters. It is difficult to answer questions about human behaviour separately, chiefly because any natural disaster has certain unique characteristics. In the following pages an attempt has been made to address these issues in the context of the 2004 tsunami. Although some relevant queries and their probable answers have elsewhere been addressed by the present author,[1] some other issues need to be investigated in the wider context of natural disasters. These have a rather different overtone: the socio-economic perspective of natural disasters can scarcely be overlooked. The social aspect should be prioritized because so far there has been an obsessive preoccupation with the economic implications of natural disasters in a globalized world. The IDNDR and the WCDR have been largely responsible for this. The WCDR has of course deliberated on issues like poverty reduction, vulnerability, development, etc. which have special economic and social significance for developing countries. Anti-globalist views (particularly those held by Joseph Stiglitz) and pro-globalist views (like those of Jagdish Bhagwati) can be readily cited.

But we refrain from citing these views because of the disjuncture between the context of natural hazards and post-disaster activities. Although there has been a host of studies and research prior to the IDNDR, the most notable seems to have been in the early 1980s where the subject of discussion was the relationship between natural disasters and development in a rapidly changing world. In1983, Kenneth Hewitt edited the collection of essays entitled *Interpretations of Calamity*. It was perhaps, the first attempt to interpret disasters from the Marxist standpoint. Ulrich Beck's book *Risk Society: Towards a New Modernity* (Sage, London, 1992) treats the diversity of risk in order to understand globalization and the tendency to think of disasters 'as though they are located outside society'. The *World Disaster Report* (2001) compares data on extreme natural events in countries with high, medium and low scores on the Human Development Index. According to this report 2,557 natural disasters have occurred between 1991 and 2000. The excellent study of the level of human development and disaster impacts was taken several steps further by the UNDP. In 2002, the UNDP commissioned a project to study mathematically more than 200 indicators of disaster risk vulnerability included in the *World Vulnerability Report*. The Human Development Index has emerged as the best instrument for predicting deaths due to extreme natural events on a global basis for an average of 20 years (1980-99). One of the principal paradigms of risk is the non-linear relationship between modern development and risk. An important associated issue is the potential impact of climatic change on individual economic sectors, particularly agriculture. P Susman et al argue that material security and prosperity for some were bought at the systemic production of greater risk exposure for others.[2] It appears to some that disaster aid at the global level is often aimed at maintaining the status quo. Cynical as this may sound, it may be the basis of underdevelopment, and it thus ensures the cycle of disaster and underdevelopment. Beck's thesis on 'risk society' has several limitations and therefore, cannot be treated as the last word on economic globalization. The neo-Marxist perspective, according to Hewitt, is to look upon natural disaster as inextricably linked with social structures. He apprehends that policies on preparedness might not be immune from struggles for acquisition of resources and polities, thereby distorting the risk from disasters. The opposite point of view refers to the limitation of this analysis, mainly its inability to identify the importance of individual agency in the 'production of vulnerability'.

The much talked of need for an early warning system in the Indian Ocean is not confined to discussions among the global leadership only. The inhabitants of the the slopes of Mt Merapi in central Java are aware that the world's most active volcano might erupt any day, and live in constant fearful anticipation of the event. An early warning system is imperative for these communities and they also need to be sensitized about the consequences of such a disaster. They are aware of the forbidden zone where they are not supposed to have settlements and they also know when to be alert or begin to evacuate. Unfortunately, no such knowledge systems existed for the Acehnese in Meulaboh, Banda Aceh and other nearby towns and villages hit by the earthquake and tsunami in December 2004. While massive tsunami waves could be detected by the Pacific Tsunami Warning Centre in Honolulu these were meant for countries in the Pacific Rim. It is also reported that some countries bordering the Indian Ocean received the warning on Sunday morning but could not respond to it because they lacked the resources for dealing with such a crisis. Experts often draw comparisons with similar past incidents where there was a massive loss of lives and property, e.g. the eruption of Mt Krakatoa off the coast of Java in 1983 caused about 36,500 deaths across the Indian and Pacific Oceans. According to National Geographic News, the tsunami in 1782 that was caused by an earthquake in the South China Sea and killed 40,000 may be considered as a comparable predecessor of the December 2004 tsunami. However, the Indonesian community still believes that an early warning system would have been of little help in Meulaboh, because of its proximity to the earthquake's epicentre. As the waves hit the city just minutes after the tremor, the people there hardly had time to move elsewhere. However, people in Banda Aceh had 40 minutes for evacuation, while the people of ten other countries had at least one hour to move to security and safety. Why were there no definite policies to alert people to the vulnerability of the coastal environment they lived in? Why couldn't they follow the procedures used by the inhabitants of Mt Merapi for combating disasters? Although people on Mt Merapi have not been formally sensitized, past experience has led them to devise their own precautionary measures. The alarming widespread indifference has prompted some people to wonder to whom the work on warning systems should be entrusted. While such policies may be initiated by a global leader or a conglomerate of countries in a particular region, its operations in subsequent years may lack the bold inventiveness of the incipient idea.

The tremor (which was the first indication of an undersea quake)occurred at 9.40 pm and it took, reportedly, one hour and 14 minutes to pass on the information to the crisis management group of the relevant authority. It took a further 36 minutes to issue a tsunami warning to the vulnerable state establishments. People have questioned the time lag between the occurrence of the event and the receipt of the information by the state authority. The latter would have taken some time to initiate the post-disaster operations, but it was certainly too late by then. Insufficient information may have been passed from one level to another. But in the current age of IT, it is terribly unfortunate that information could not be conveyed fast enough. People are skeptical of the hierarchy or bureaucracy that prevents them from acting on an emergency basis—they are seldom allowed to act on their instincts and rise to the occasion. Why shouldn't technology be used to empower people in potential disaster areas to receive information from the original sources in the least possible time? A Houston company claims to be able to detect tsunamis with the instruments and equipment that it manufactures. According to the *Houston Chronicle*, Oyo Geo Space (known for its oilfield products) is one of the concerns developing the tsunami early warning system for the Indian Ocean. Perhaps establishments like the Oil and Natural Gas Commission (ONGC) of India could explore the possibilities of partnerships with concerns like Oyo Geo Space.

Widespread destruction caused by natural disasters results in a media explosion. News about losses due to disasters pours in from a variety of sources. Broadly speaking, these sources include reports in the major media forms (print and electronic), government releases and a host of miscellaneous publications prepared immediately after a disaster. Later, these are consolidated for publication in disaster sourcebooks, encyclopedias and a wide range of research publications, bulletins and works by professional emergency managers, NGOs and other groups. There are also many international bodies and institutions that have data banks about major types of natural disasters. The following are glimpses of media accounts of the recent tsunami which helped to develop an initial information base.

The body count continued for days after the tsunami and the death toll mounted steadily across Asia. On 30 December 2004, Sri Lanka had revised its toll to 27,268. The combined death toll in Thailand, Malaysia, Somalia and elsewhere, ravaged by giant waves, was expected to run into thousands. The Indian toll, at the time was about 7,000, but unofficially it was much higher at 15,000 or more. Apparently as many as 5,000 foreign tourists were killed. The official toll on 2 January 2005 was: 80,246 in Indonesia, 28,729 in Sri Lanka, 4,812 in Thailand and 90 in Myanmar. The eastern parts of Sri Lanka experienced flash floods on 1 January 2005, forcing the evacuation of about 2,000 people. According to an official of the Natural Disaster Management Centre at Sri Lanka, 16,525 people were injured and 5,240 were missing.

In Indonesia, Sumatra seemed to be the hardest hit area. Heavy rains on the corpse-littered Indonesian city of Banda Aceh on 1 January created conditions for diseases such as cholera, diarrhoea, etc. The tsunami spared only a handful of buildings in Banda Aceh. According to contemporary reports, about a quarter of the town's 40,000 people were feared dead but only a fraction of that number could be found. Heavy rain prevented relief workers from collecting corpses. The incoming figures kept varying: on 4 January the death toll in India rose to 9,479, while 641,072 people had been moved to safer places and another 384,956 lodged in 560 relief camps in the states affected by the tsunami. In the Andaman and Nicobar Islands the figures on 3 January were as follows: 818 (mainly fishermen) dead, 5,681 missing, 9,136 evacuated and 13,320 lodged in the five relief camps. In Andhra

Pradesh there were 106 deaths, seven were missing and 33,864 were evacuated to safer areas. In Kerala there were 166 deaths, two were missing, 55,557 were evacuated and 5,557 were lodged in relief camps. In Tamil Nadu the death toll was 1,400 on the evening of 27 December 2004, including 125 deaths in Chennai; nearly 700 lives were lost at Nagapattinam, 200 at Cuddalore and 250 at Kanyakumari. In Pondicherry the initial death toll of 102 continued to rise steadily. In Andhra Pradesh the death toll was 61; this included 24 deaths in the Krishna district, seven in Nellore and one in Guntur. 200 fishermen were reported missing in the East Godavari district.

During the time lag between a natural disaster and the arrival of external aid, resources available in the immediate neighbourhood of the disaster can be utilized to control the situation. The situation and particularly human suffering is likely to be exacerbated if relief measures are unnecessarily delayed. In the wake of a natural disaster there are often instances of camaraderie between countries with sensitive and unresolved political issues between them. Both Indian and Pakistani air force planes and naval ships rushed to Sri Lanka and the Maldives, while US air force planes landed in Banda Aceh and Thailand with aid. Obviously geopolitical problems were set aside and the relative distances between countries ceased to be a prohibitive factor. Certain factors did, however, come in the way, e.g. problems of access to the disaster site and conflicting efforts by relief parties. Also, communication networks were clogged and information was unstructured, uncoordinated and ineptly managed in coastal areas affected by the disaster.

This revealed the disturbing inadequacy of knowledge or recognition of available on-site resources. Alternative emergency resources could hardly be identified and were thus inefficiently mobilized. Surprisingly, the indigenous population of the Andaman and Nicobar Islands handled the situation fairly well with their apparently primitive wisdom and practices. The Andaman airport was in a shambles. In addition to a natural disaster emergency plan, an island airport emergency plan is necessary to reduce the loss of lives and avoid the disruption of stable aircraft operation. Quite obviously, special infrastructural problems are posed by an island with a single run-way. In such cases there is hardly any option but to fall back upon the island's own resources until the situation improves and the island becomes more accessible. Two crucial questions need to be addressed. On the first impact of a natural disaster how can the administration utilize available resources? Can it afford to augment the available resources?

The predicament in Sri Lanka and the Maldives was horrifying. At the southern fishing village of Mirissa, where 300 boats support some 1,300 families, only three boats of fishermen could return to sea at night and send part of their catch north to the capital Colombo according to reports of relief workers. Many of the fishermen whose boats had survived the catastrophe still refrained from going out to sea (at the cost of their livelihood) because of a local superstition that a deadly virus had attacked the sea shore. The effects of the tsunami were visible to those who flew over the destroyed areas of Banda Aceh, Indonesia. The tsunami swept away everything in its path—houses, ships, cars, etc.—only stopping where it washed up against the city's foothill. The UN estimated 1.5 million casualties across Asia. It was a pity that even during such a crisis some countries fought to out-manoeuvre others in terms of toil and assistance. New Delhi's stance, on the contrary, was most genuine and it despatched three service ships converted to hospitals-at-sea to Nicobar, Sri Lanka and Banda Aceh.

From an economic point of view, it seemed that the initial economic toll would be US $14 billion. Economists started revising their negative growth estimates. Munich Re, the world's largest re-insurer, has estimated that the damage caused by the tsunami cost more than US $13.6 billion (i.e. Euro 10 billion). The ANZ Bank's economists said: 'There is no close correlation between the magnitude of the disasters in terms of loss of life covered and to damage in financial terms.'[3] David Cohen of Action Economics in Singapore felt that: 'What clearly distinguishes this disaster from the rest is the magnitude of the death toll . . . most of the damage, aside from the death toll of course, was in the residences and it is a heavy price for the people but it won't have that much of a subtraction from production capacity, the exception being the tourism industry in Thailand.'[4] Thailand, which receives its tourism revenue from Phuket and Krabi provinces suffered heavy losses. Standard Chartered Bank estimated that the GDP in Sri Lanka and the Maldives would be worst hit; the GDP for 2005 in the two countries would be reduced by 2 per cent and 4 per cent respectively.[5] According to the Standard Chartered report the scale of this disaster is far greater and it has impacted more countries than earlier natural hazards. The report states that even though there may be a huge economic setback in the beginning, one will probably witness a strong policy response and economic recovery. The fact remains that while there would certainly be a terrible economic impact on the fishing industry and tourism, one would hardly be able to quantify the direct economic impact. Asia's economies can rely in part on foreign currency reserves, but there have to be other sources of relief and reconstruction as well. Indonesia has invested US $1 billion to rebuild Sumatra, while Thailand has invested US $510 million to rebuild its tourism industry.

The financial sector is essentially composed of insurance, banking and other concerns. Past figures support the general consensus that disastrous events have escalated exponentially and the resulting economic and insurance losses have increased manifold. The tsunami has drawn attention to broad insurance issues. The western world had considered a variety of insurance models, but what could be an appropriate insurance model for a developing country like India? How are Indians to adapt such models for other uses? These concerns ought to address the potentially enormous losses likely to result from natural disasters. The scientific investigation of natural disasters has not been neglected. Insurers may seek to alleviate the problem of future losses by investing in alternative energy technologies. Given the global interaction between developed and developing countries on matters of energy security, one can reasonably posit similar initiatives between developing countries in South East Asia. Indeed, the insurance industry has an important role to play because of its immense market leverage.

Profiles on responses to natural disasters ought to focus on the behavioural responses of children, usually a sorely neglected subject. The UN has insisted that orphaned children be registered and placed in the care of a responsible adult. In many disaster-affected areas the UNICEF has found to its dismay that orphans often become victims of child trafficking; hence their registration is of paramount importance. The children at a centre for physically disabled and mentally ill children at Galle were lying helplessly in their beds as water entered their dormitories during the tsunami. Some desperate children managed to move to a higher place as the water level rose inside the one-storey *sambodhi* shelter; others were swept away. Only 41 of the shelter's 102 residents survived. The caretaker of the home was stunned to see the sea rushing inland faster than a speeding car. Children screamed and many were dumbfounded at the sight. They were told to exit the building immediately, but many of them could not understand the instructions or the danger that they were in. A woman was frantic when she found herself separated from her son, but later found him lying on the beach after the waves had subsided. The child could be revived only after water was ejected through his nose and mouth. The severe shortage of food and supplies drove people at an evacuation camp in Banda Aceh to eat uncooked noodles. The Medan-based Aceh Sepakat Foundation reported that at least 20 Acehnese children were thought to have been smuggled out of Aceh after the 26 December calamity. However, steps were taken by the government for maintaining surveillance at the airport and seaport in Aceh. Indonesia and Malaysia were not the only countries in South East Asia where such atrocities were perpetrated. There were also memorable incidents of heroism and courage, for example a 12-year old girl called Meghna, reportedly the daughter of a meteorological officer stationed at Nicobar, swam and walked for two days to reach Warrangal in Andhra Pradesh after losing her parents to the killer wave. According to her, when the waves struck their home they rushed up the stairs to the second floor. Her parents and brother were washed away. She plunged into the waves and swam and finally walked over patches of dry land and then swam again. Psychiatrists ought to tend to survivors with post-traumatic stress disorder and orphaned children must be rehabilitated with the necessary psychological aid. At the global level, the WHO regional office for South East Asia has been working with other organizations in the region to train community workers to counsel the traumatized survivors. Mental health experts could diagnose the emotional trauma suffered by the victims of the Indian Ocean tsunami. According to the WHO regional director for South East Asia, 'The mental anguish of those who narrowly escaped the tsunami or lost close relatives is made far worse because many of them have lost their homes, their jobs and their possessions. These people urgently need support . . . Many can cope and will gradually come to terms with what has happened but others will either sit motionless or cry for hours on end. If support is not urgently provided, the long-term effect on the population could be terrible.' The WHO India office has provided psycho-social support to people in the tsunami-affected areas of India on a priority basis. This office, in collaboration with UNICEF and other centres helps to coordinate the efforts of local health workers, and tries to ensure the broadest possible outreach of their endeavours. WHO has sent mental health experts to Sri Lanka and Indonesia to work with local mental health professionals. In Thailand, a WHO official observed: 'Some of these people (victims) are literally suffering a second tsunami. Today we saw a son, who was extremely distressed at the guilt he feels for not having been able to save his father from the force of the tsunami. Such people urgently need a support system to help them cope with this traumatic experience. If such support is not provided, it could very likely end in yet another tragedy for some of the families and we simply cannot allow this to happen.' There is an urgent need for services extending beyond mere compassion or empathy for tsunami victims.

The financial impact of the tsunami is another major area of concern. Munich Re, the world's largest re-insurer has done much to help recover losses. Insurers have to take cognizance of the fact that tsunamis are mega-perils having return periods of several years. Every tsunami is foreshadowed by an earthquake, a volcanic eruption or a landslide. A question often asked in the context of the tsunami is: if tsunamis are to be treated as an insured risk, should events like earthquakes also have an insurance cover?

The immensity of the December 2004 tsunami has motivated, if not *urged* reactions from various individuals and groups. The following is a selective overview. Joseph Stiglitz, the 2001 Nobel Laureate in Economics, felt that the televisual transmission of vivid pictures of the destruction was a 'demonstration of the power of globalization'. Indeed the world truly seems like a 'global village' during times of crisis. Stiglitz was vocal about the initial 'niggardly' provision of aid by the world's richest and most developed nations. Relief efforts must necessarily be multilateral; thus he emphasized the formation of a core group among nations in the region under the leadership of the UN. According to him, the response from certain countries was remarkable. G-7 countries announced their decision to provide debt relief in response to the disaster; Stiglitz applauds the way countries like Thailand rose to the occasion. This went a long way in helping Indonesia, worst hit by the tsunami, which had to repay a debt of US $41.32 billion. Stiglitz has wisely cautioned: 'Nature, as the tsunami makes it clear, has its timetable. Unless we learn to respect it, we will all miss the boat.' He further said: 'We have to learn to cope with the forces of nature as we can't alter them!'[6]

The above remarks should not be interpreted to mean that Stiglitz did not lament the absence of a warning system in the Indian Ocean. Rather, he reflected on the possibility of developing such a system, with state-of-the-art technological inputs. He sincerely hoped that an unequivocal consensus could be reached. It would be a pity if things went the way of the Kyoto Protocol which was accepted and passed only after modifications that affected its original message.

It is interesting to examine Prof. Tadapally Murthy's comments about the tsunami warning system in the Indian Ocean. A leading scientist based in Ottawa, Canada he has visited India several times and was an active participant during the Indian Science Congress (1991) whose focal theme was 'Coping with Natural Disasters: An Integrated Approach'. He discussed how he had helped develop the Canadian Tsunami Warning System some 30 years ago. Agreeing that tsunamis affected India fairly frequently, he asserted that a venture on early warning systems would not be a futile exercise. In fact he has recently pleaded for some short-term measures, such as installing seismographs and tide gauges on the sea floor, particularly in sensitive areas of the Indian Ocean, Bay of Bengal and Arabian Sea where tsunamis occur, so that West Bengal, Orissa, Andhra Pradesh and Bangladesh could be better prepared.[7] As Prof. Murthy mentions in his interviews, commentaries and speeches, India's recent 'resource crunch' should not hinder the acquisition of equipment necessary for a tsunami warning system. He referred to the inexpensive system he had himself devised.

Reports and commentaries produced after natural disasters rarely penetrate the human element woven into disasters. While these descriptions assume transnational dimensions, the heterogeneity of the affected society is often ignored. An approach combining the transnational and the social/psychological ought to feature prominently on any agenda pertaining to tsunamis.

The use of disaster datasets is now attracting greater interest than ever before. Prof. James K Mitchell of Rutgers University, USA, wrote an excellent occasional paper on the eve of the IDNDR.[8] This remarkable paper raised a variety of issues, keeping the imperatives of the IDNDR in sight. Although the paper's chief concern was the data on losses due to disasters, the advantages and limitations of available data and global disaster losses have been brilliantly assessed by Mitchell. Neither did Mitchell overlook disasters in India, China, Bangladesh, Japan and neighbouring countries. Important losses have occurred in the Asian Rim and even the Pacific Rim has been much in the news lately because of the Indian Ocean tsunami. Given the UN's present perplexity about the final figures on losses, a study modelled on Mitchell's investigation of disaster loss datasets would be most helpful. The advantage of the GIS technology is that it not only integrates heterogeneous data but also helps to assess pre-disaster vulnerability and develop post-disaster mitigation strategies.

Awareness of natural disasters is essential at all levels. The IDNDR and WCDR have reiterated the lessons of earlier disasters. Perhaps we need to ask the same questions as Peter Drucker—what information do we need? From whom should we be getting it? What new tasks can we tackle now that we have all these data? Which old tasks should we abandon? Which tasks should we do differently? According to Drucker, a database contains the 'ore' of information, rather than information itself. The conversion of the 'raw' facts of natural disasters into meaningful information involves several phases and careful decisions. One needs to consider whether the information is reliable and adequate and a principle of selection ought to prevent an information overload. The processing and dissemination of data is sometimes referred to as knowledge management. Disaster literacy is of paramount importance in the twenty-first century. It is a prerequisite for what is often called an 'information society'. It ought to be considered a part of the 'basic human right of lifelong learning', to borrow a phrase from the Prague Declaration, 'Towards an Information Literate Society' (2003). In collaboration with each other, South East Asian countries ought to create teams of disaster literate activists, whose skills and knowledge are indispensable to post-disaster programmes. Since natural disasters undermine and unsettle global collaboration and cooperation, IT Enabled Services (ITES) could help tackle emergencies and crises. This section may be concluded by stating that the flow of information is an integral constituent of development. The notion of development in turn ought to be treated as a vade mecum for the WCDR. Information literacy must be accorded a higher status on the agenda of programmes and activities. The participation of all segments of society, and especially the educational sector, is essential.

NOTES

1. Sinha, D K, 'Whither Contours of Natural Disaster Management?', Special Volume on Disaster Management, *Asian Studies*, No. 2, 2002, pp.2-6.
2. Susman, P, O'Keefe, P and Wisner, B, 'Global Disasters: A Radical Interpretation' in Hewitt, K, ed, *Interpretations of Calamity: Viewpoints from the Perspective of Human Ecology*, Allen and Unwin, London, 1983.
3. *The Times of India*, Kolkata, 31 December 2004.
4. Ibid.
5. Ibid.
6. 'Lessons from the tsunami', *Economic Times*, Kolkata, 9 February 2005.
7. For a brief account of Prof. Murthy's views, see Radhika, V, and Ali, Farwa Imam, 'Track a tsunami', *The Week*, 9 January 2005, pp.36-7.
8. Mitchell, James K, 'Where might the International Decade for Disaster Reduction concentrate its activity: A comparative analysis of disaster datasets', The Natural Hazards Research and Applications Information Centre, 1989.

Chapter 9

Global Support, Local Settings

It is widely agreed that any contemporary event or phenomenon has not only local but also global consequences. The relation between the local and the global must be recognized for efficient disaster mitigation measures to become operational. Active participation from experts, field workers and local and international activists would greatly assist disaster relief efforts and the IDNDR, the WCDR and some other recent summits held after the tsunami have established the broad patterns of global disaster mitigation efforts. Disaster risks and vulnerability can be checked only with the help of certain key elements of globalization. Social diversity and complexity ought to be considered and the latter must be seen as complementary to the local socio-cultural milieu.

Vulnerability is often regarded as a classic symptom of despair. Recent natural disasters have helped to shape international relations and international cooperation seems to have soothed some of the conflicts among hostile countries or states. It is not that recourse is taken to so-called 'global interventions' under the label of disaster diplomacy. Internationally acceptable perspectives and aspirations must be identified and related to local milieus. An understanding of local adaptive capacities introduces us to community-based vulnerability. Risk patterns are being globally recognized and negotiated. While many areas of concern have been identified, human adaptation is possible in very few cases. Thus disaster mitigation has not improved substantially.

Having dwelt on global and local trends, one may consider a perspective that transcends both. Often described as a 'meta-trend' it does not reject the essence of either trend, but tries to achieve a broader and composite perspective. If a country is amenable to globalization, it certainly has an edge over other countries. According to Thomas L Friedman, 'That's globalizing at its best.'[1] South East Asian culture is particularly complex. Over centuries, countries in the region have absorbed ideas and practices from others and blended them with their own traditions. Although some of these countries have been unable to participate significantly in globalization, they have fostered 'adaptability' and 'adoptability' among themselves. Ultimately, it is the richness of a country's culture that defines its status. But culture is context-specific, leaderships change and the degree of adaptation varies. Globalization thus evolves with cultural changes.

The local vulnerability of a region like the Philippines, which is frequently lashed by typhoons and tsunamis, is obvious. In the context of the Philippines, the word 'community' is often used to describe the geographically and politically defined unit of the barangay. Local leadership is provided by each barangay. The Barangay Disaster Action Team (BDAT) has worked for a variety of mitigation measures. The Integrated Community Disaster Planning Programme (ICDPP) in collaboration with the PNRC has built links with barangays at the community level to discuss aspects of social vulnerability related to typhoons. A collaboration of this kind allows the reduction of vulnerability with the help of local knowledge, experience and capability. While it is not easy to forge links with global organizations, this is amply compensated for by the communitarian approach which is predominantly local but not insular enough to lack global characteristics.

The other strand of global perspectives emerges from what is often described as political ecology. Here the adjective 'political' is not ideologically loaded because of the primary emphasis on *ecological* systems. Some commentators have even described globalization in terms of global localization. Countries are often tempted to develop both global and local political relationships. What results is often described as asymmetric interdependence. The asymmetry arises due to interdependence and connectivity, which often tend to be stronger in some ways than the other elements. This applies directly to natural disasters if there are interfaces with ecological systems. The ecological framework becomes an imperative for global actions. A local framework must comprise elements of the local ethos and culture, and must include elements of heritage, equity and security. Obviously, the last two elements are predominantly socio-economic and thus, vulnerable to social and economic factors. In the immediate aftermath of a disaster it is often forgotten that relief involves much more than the mere distribution of money. The inability of fishermen to retrieve their earlier livelihoods after the tsunami is evidence of social vulnerability in the wake of a natural disaster. A fruitful partnership of global and local aid is possible only if local systems are receptive enough to absorb global support. Local efforts will find it difficult to combat natural disasters without global aid. Technology and the transfer of knowledge can serve as essential links between the two levels.

There were numerous cases of shock and trauma after the tsunami. Local community workers, who are aware of the needs of disaster victims, should be sensitized if not formally trained in psycho-social support. The relief activities of the WHO in the affected countries along with counsellors from the region, familiar with local culture, demonstrated how a combination of global expertise and local experience could help restore the psycho-social behaviour of the endangered community.

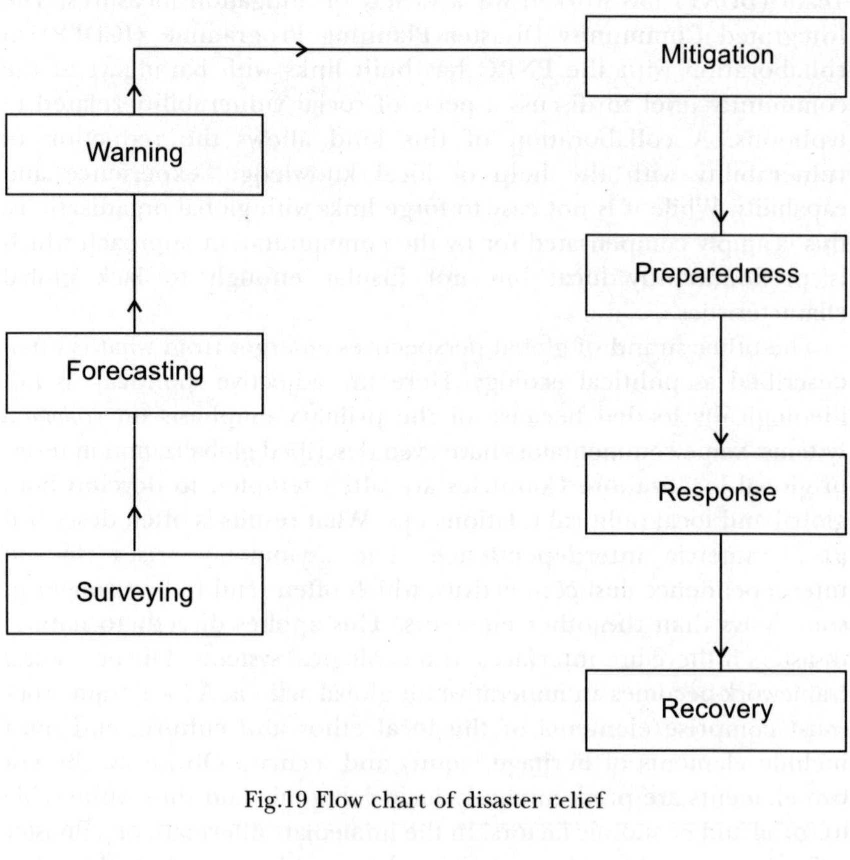

Fig.19 Flow chart of disaster relief

NOTES

1. Friedman, Thomas L, *The World is Flat: A Brief History of the Globalized World in the Twenty-first Century*, Allen Lane, London, 2005.

Epilogue

A RETROSPECTIVE: SEEKING ANALOGUES

The foregoing chapters have outlined the influence of the WCDR and the Hyogo Declaration on disaster mitigation, not just in South East Asia but also globally. The flow of aid and relief that followed the 2004 tsunami is still fresh in people's minds. Natural disaster management has received fillip from new sources and sectors and the possibility of introducing or strengthening insurance or even re-insurance is being explored, particularly in countries where financial support for relief and recovery has been in the nature of subsidies. This affects the budgetary allocations for natural calamities. Further, with the liberalization of the economy, there is now a tendency to seek greater involvement of the corporate sector. Of course, there have always been large corporate donors. Perhaps for the first time, various business chambers like the Federation of Indian Chambers of Commerce and Industry have involved themselves on a very large scale. Until recently, many private sector disaster management initiatives were developed to minimize property damage and loss, but lately the private sector has begun rethinking its roles and obligations. The sector's concern with contingency planning was quite natural, considering that the disaster affected their business interests. Corporate sector intervention can easily be attributed to the developments in communication and other advances in technology. Also, economic development in South East Asia has brought tremendous social benefits to the countries in the region. Governments increasingly recognize that risk management is the key to continuing economic development; the corporate sector too is now aware that risk management is an important prerequisite for corporate transactions. The Hyogo Declaration strongly emphasized that developmental processes need to be reckoned with while dealing with natural disasters. There may be some hesitation about repositioning the corporate sector to play a role in preventing and mitigating hazards, but both natural disasters and technological hazards actually share much in common with respect to the process of prevention and mitigation, education and training, damage control and public awareness. Most elements of risk management used for technological disasters may be applied to natural disasters as well. All governments and political leaderships are well aware of the consequences of natural disasters and have, to some degree, managed such disasters on certain occasions. Aware of the implications of the failure of efficient management, they realize the importance of partnerships between the government, public sector and corporate sector. NGOs also supplement disaster management teams. The losses caused by natural disasters cannot be assessed in purely economic terms. As natural and technological disasters result in huge losses for businesses, the latter must not merely *envisage* disasters but work actively to *prevent* them.

A private sector summit was held on 12 May 2005 to raise funds for tsunami victims. The corporate leaders emphasized the need for foreign aid, not only for post-tsunami reconstruction but also for long-term growth. The US Chamber of Commerce, the US ASEAN Business Council, The Asia Foundation and the Asia Society sponsored the day-long 'Private Sector Summit on Post-Tsunami Reconstruction and Rehabilitation'. It provided a platform for corporate representatives, senior government officials from areas badly affected by disasters and representatives from NGOs to deliberate on ways and means for the private sector to extend help and support for reconstruction efforts. The private sector was urged to continue providing (preferably financial) support and other types of assistance. Many of the citizens of the countries struck by tsunamis had suffered a violent psychological shock; thus humanitarian efforts had to be made to help restore the people's resilience. The summit demonstrated the uninhibited collaboration of the corporate sector. It is thus possible for a new kind of disaster management team, continually developing its crisis response capabilities, to emerge as a non-doctrinal support force.

Disaster awareness appears to develop differently in different disaster situations, chiefly because of the uniqueness of each disaster. The degree of difference in the perception of risks posed by disasters is remarkable. For each affected individual there exists a complex of relations between his/her anticipation of or ideas about natural disasters and the event itself. Knowledge of the victims' psychology is important for relief workers; this is an aspect that the latter tend to overlook in the course of their work. Unless relief workers consciously acquire this knowledge *in situ*, their efficiency during crises will diminish. The choice of post-disaster procedures is a vital element in mitigating the impact of disasters. Of course the particular context of each disaster determines the procedures to be followed. The resistance shown by the indigenous populations of the tsunami-ravaged Andaman and Nicobar Islands can be comprehended only as a complex of many factors. Unless people are alerted to the multiple facets of natural hazards or made aware of case studies of particular disasters, they will be taken by surprise when a disaster actually strikes. The framework of facts and hypotheses about a community's collective beliefs, ideas and experiences constitute the 'experiential context' of a natural disaster. While the cultural similarities and diversity of South East Asian countries need to be identified, cultural conflicts should also be addressed in order to mobilize efficient response mechanisms. Of course, a caveat that needs to be appended is that 'culture' simplifies, standardizes and idealizes in order to reflect a uniform reality.

The WCDR and the Hyogo Declaration have taken an unequivocal stand with respect to poverty alleviation and development, which are inextricably linked with advances in technology and the diffusion of 'knowledge'. Although the phrase 'global village' itself does not recur very frequently in the text of the declarations, the connotations of the phrase are strongly felt throughout.

Where do we go from here? How can the workflow be better channeled? How can the directives of the WCDR be better implemented? Perhaps the transfer of aid from developed to developing countries is necessary. Can a framework (perhaps a consortium of South East Asian countries) be created to stimulate efforts on natural disaster reduction and facilitate innovation for human resources and institutions in the constituent countries? How can human capital and technological diffusion complement each other? How can the inter-flows and intra-flows between and among South East Asian and Western countries catalyze the development envisaged by the WCDR? How can developmental problems and processes be formulated? The potential and utility of the stock of human capital is limited; even the mobilization of resources from elsewhere (at a distance from the disaster zone) has its limitations. We must not only identify the characteristics of a disaster in a particular region, but also develop an anticipatory understanding of what a disaster brings in its wake.

It must be reiterated that 'development' is the cornerstone of any blueprint emerging from the WCDR and the Hyogo Declaration. 'Knowledge,' 'knowledge economy' and allied ideas are all concomitants of development. The triad—'information,' 'technology' and 'development'—must necessarily be taken into account. The long period of indifference to and inaction about problems of global development ended abruptly with the tsunami in 2004. Almost overnight the global community began to focus its strategic attention upon local communities in order to help them escape the traps of poverty and allied social and economic ailments. That was the rationale of the UN Millennium Summit, whose finale was in September 2005. As the UN Secretary General Kofi Annan pointed out, new technology offers an unprecedented chance for developing countries to leapfrog earlier stages of development; everything must be done to maximize peoples' access to a new information network. It follows readily that countries unable to 'get connected' will lag further and further behind—that is one of the unfortunate results of a global economy. Unfortunately, the digital divide has widened in the last couple of decades. How significant and serious is the divide for developing countries? How is the gap to be bridged? How can the versatility of ICT be utilized for development, especially since ICT transcends geographical barriers? The role of ICT in development strategies should be strengthened even if it appears to contribute little to the more obvious development goals. These challenges are presented by the UN's Millennium Development Goals (MDG). During the IDNDR, the impact of technology was marginal. New technology is hardly an end in itself; its contribution to the overall development strategy is important. Powerful new tools (products of technological advancement) need to be made available to potential stakeholders.

Early warning systems, forecasts and preventive measures are being renovated and re-installed on the basis of newly acquired technology. The seismic resistant or detection systems—whether in the Yangtze in China, the Mekong in Vietnam or the Ganga-Brahmaputra in India—are far more sophisticated than they used to be a decade earlier. The IDNDR might have reinforced activities and programmes to contend with possible disasters. But it is debatable how many of these projects are digitally empowered. There is still a wide divide between developed and developing countries. The means to equip the latter with the necessary technological know-how needs to be addressed. While technology has taken great strides forward, vast segments of the population continue to suffer because of unforeseen natural disasters. Natural disasters like tsunamis will continue to present a grim vision of the future unless the 'gaps' in development are closed with innovative new technology.

'Development' as envisaged by the Hyogo declaration, should not be understood in its conventional sense. The scale of development ought to embrace the community welfare of a particular region. Although global consortia conduct regular disaster mitigation programmes, every constituent nation of the South East Asian consortium has limitations with regard to disaster-related services. Hence, there must be a flow of resources from the global to the local. The expansion of digital networks is often considered to be something of a panacea in such contexts. High-capacity fibre-optic links span oceans, crisscrossing China and other countries. It is a myth that protectionism will be jeopardized if poor communities are better connected to the world around them. It is only rational to grant farmers access to weather forecasts which would enable them to plan their harvest and take the necessary measures for a higher yield of crop. Greater connectivity among South East Asian nations will create new international opportunities and also alert geographically isolated communities to the destructive powers of natural disasters. Obviously, the WCDR and/or the Hyogo Declaration must have a self-sustainable model of development. The new development model posits interaction with earlier processes and an adherence to the values of the MDG (adopted at the UN summit in September 2005). These countries, when digitally empowered, can envisage both, horizontally and vertically, a triple convergence of development, technology and their cultural ethos. Every individual is rooted in a particular historical moment, shaped by tradition, culture, global and local factors. Dialogue between nations can be traced to antiquity—this tradition of interaction ought to stimulate deliberations on the mitigation of natural disasters.

SITUATING NATURAL DISASTERS IN THE SOUTH EAST ASIAN DOMAIN

The two post-tsunami natural disasters that have generated the most concern and interest are the recent Hurricanes Katrina (in New Orleans, USA) and Rita (in the Gulf of Mexico). The media has not presented a very clear picture of how natural disasters have been handled in the past. The communities directly affected by disasters have learnt to live with them, struggling for survival and the protection of property and even the corporate sector did not hesitate to capitalize on the destruction caused by a hurricane. As Thomas L Friedman points out, 'During hurricanes, Wal-Mart officials told me, Wal-Mart knows that the people eat more things like Pop-Tarts—easy to store, non-perishable items—and their stores also sell a lot of kids' games that don't require electricity and can substitute for TV. It also knows that when hurricanes are coming, people tend to drink more beer. So the minute Wal-Mart's meteorologists tell the headquarters a hurricane is bearing down on Florida, its supply chain automatically adjusts to a hurricane mix in the Florida stores, more beer early, more Pop-Tarts later!'[1]

Disaster management is yet to be undertaken in remote areas, even by the corporate sector. The latter ought to extend to natural disaster management the same improvisatory techniques that it applies to its global supply chain. A new method of supply chain management is necessary to relocate evacuees efficiently. Friedman, in his popular book, makes full use of this mode of horizontal collaboration and creation through global supply-chaining which is, in a way, the essence of the 'package flow of advances in technology'. A situation-conscious disaster manager can optimize the total work involved using a mathematical model. The network of South East Asian countries may enforce such a supply chain mechanism to facilitate the flow of global efforts.

Resources and search mechanisms must unite communities in a common cause and individuals and communities must assimilate the latest advances in technology. If we are to understand how a disaster is potentially situated, we must actively involve people who help to prevent natural hazards and sensitize communities to them. This in turn requires skills for collaborating horizontally and managing horizontally, rather than following conventional top-down approaches. It is only with the implementation of such collaboration that ingenious new methods can be devised. The latest ideas about early warning systems and the recommendation of several summits in Indonesia, Phuket and elsewhere clearly show that there must be multiple convergences of the three Rs (rehabilitation, recovery and renovation) allied to natural disasters. The implosion of globalization in South East Asia has its pros and cons. A horizontal variant of collaboration must be worked out. As the Hyogo Declaration broadly indicates, 'knowledge' workers must be identified who can move horizontally into new knowledge-oriented jobs. Concept-driven programmes and activities could help to minimize false distinctions between digitally empowered work and manual operations. This is vital for long-term development. Neither is there any room for complacence: countries cannot afford to be caught unawares as they were by the 2004 tsunami.

The historical and geographical overview of South East Asian countries (Chapter 1) shows that one cannot accord preferential treatment to a particular region prone to a particular kind of natural hazard. Equal importance must be attached to disaster mitigation in different affected areas. The fact that relief operations are easier to conduct in certain regions is hardly relevant. A typhoon-ravaged area in China may be more accessible than areas hit by typhoons in Cambodia or Taiwan. But relief efforts and aid must be equally directed to these affected areas, irrespective of the ease or difficulty of setting up warning systems and extending aid.

The MDG envisage effective and durable alliances among South East Asian nations. A platform must be created for them to share their concerns and aspirations. The cooperation and potential benefits ensured by such a platform will greatly aid the process of natural disaster management. If each nation contributes to disaster mitigation, the whole will certainly be greater than the sum of its parts. The flow of benefits to participating nations will inevitably be steadier and more powerful if it is the outcome of an international corpus of aid, resources and expertise.

NOTE

1. Friedman, Thomas L, *The World is Flat: A Brief History of the Globalized World in the Twenty-first Century*, Allen Lane, London, 2005, p.136.

Appendix

A Note on Damage Assessment

The facts, figures and arguments of the preceding chapters emphasize the global character of natural disasters. The extent of damage determines the volume of aid flow for post-disaster reconstruction. This section deals with the assessment and monetary quantification of damage.

On 7 January 2005, *The Economic Times* (Kolkata) carried a report stating that the Central Government had estimated losses of Rs 5,322 crore; figures were also provided for the different states and Union Territories affected by the December 2004 tsunami. The international commitment for disaster relief was about US $4 billion, with affluent countries declaring a freeze on the debts of the tsunami-ravaged nations. While it is unfortunately true that such figures are sometimes the product of vested interests and careful manipulation, the questions that intrigue most people remain: how are these estimates made, and how are figures computed within a few hours of a disaster? Are any formulae or standards used to assess damage and loss? Or is it a purely ad hoc and empirical exercise, tempered by bureaucratic processes?

A rather naïve and simplistic approach is to begin with an average figure, e.g. of half a dozen TCs in the Bay of Bengal and Arabian Sea, perhaps two or three are particularly severe. But such a premise usually minimizes the actual extent of damage. While satellite pictures are helpful indicators of the extent of damage, one may broadly classify damage operations into: (a) examination of buildings/constructions (b) on-site hazard identification (c) methods of effecting temporary repairs. After the devastating Hanshein Awaji earthquake Japan, under the aegis of the UNCRD, worked out figures of casualties, damage to buildings, the damage of lifelines, large-, small- and medium-sized concerns, and arrived at the total cost of restoration and reconstruction. In the aftermath of a natural disaster, the establishment in most developed countries follows the same procedure.

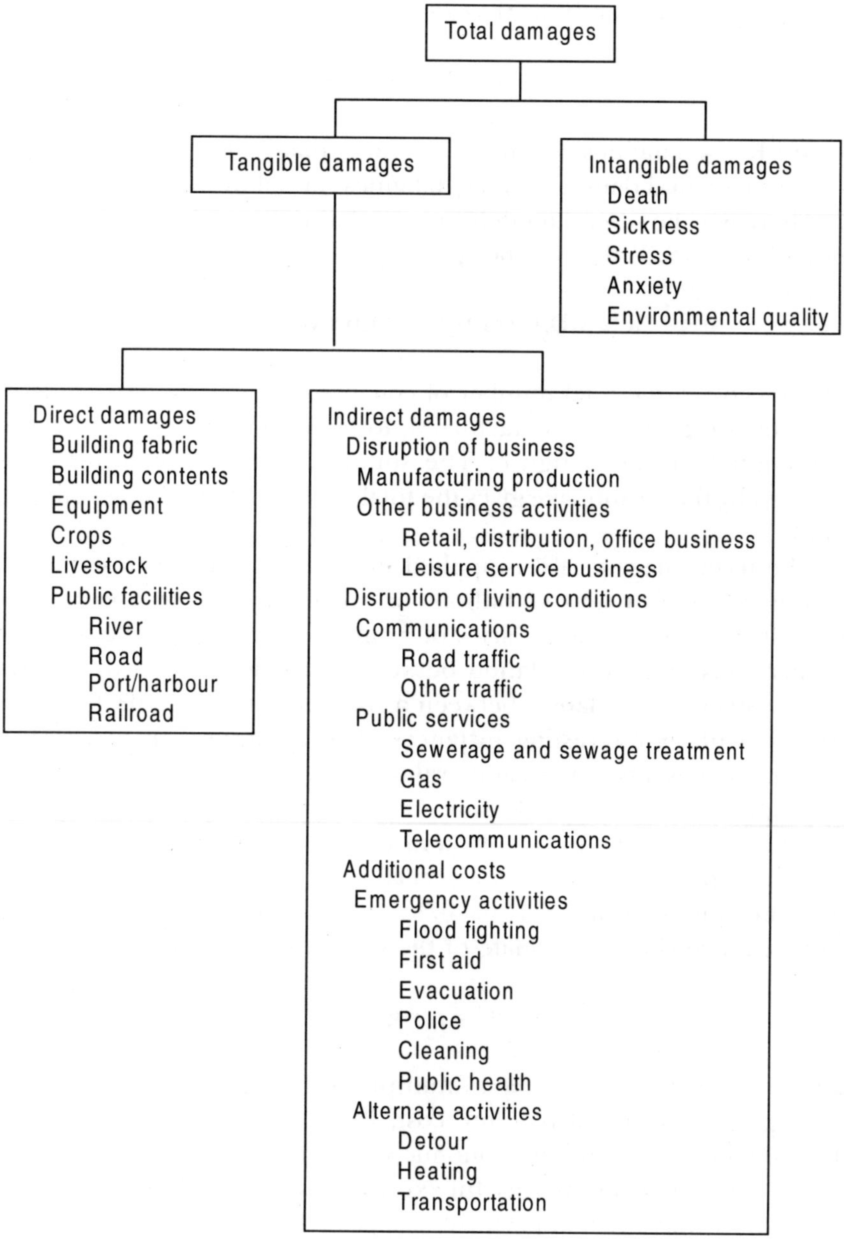

Fig.20 Damage categories

Mathematically, a generalized casualty equation has been derived to calculate the number of deaths caused by an earthquake. The equation is given by:

$$K = K_s + K_1 + K_2$$

where $K_s \Rightarrow$ fatalities due to structural damages; $K_1 \Rightarrow$ fatalities due to non-structural damages; $K_2 \Rightarrow$ fatalities caused by follow-on hazards.

There may be several models for calculating casualties, e.g. in a class of buildings ('b'), the number of people killed can be expressed as:

$$K_s b = D_5 b^* [M_1 b^* M_2 b^* M_3 b^* (M_4 b + M_5 b)]$$

where $D_5 b^* \Rightarrow$ the total number of collapsed structures (damage level 5) of buildings of type 'b'; factors M_1 to $M_5 \Rightarrow$ a range of modifiers to a potential mortality figure. The summation of all building types affected by the earthquake gives the total dead due to building collapse, i.e. K_s.

The term 'mega-perils' is applied to natural hazards whose impact causes damage of tens of billions of dollars and/or whose damage extends beyond the boundaries of a single continent. While we have earlier considered the relation between a risk and a hazard, how do we investigate the relation between a risk and a mega-peril? Tsunamis are particularly interesting instances of mega-perils. Any exercise in hazard analysis, assessment and evaluation must necessarily include the appropriate analysis of risk. Infrastructural ventures in hazard-prone areas must reckon with (often unpredictable) risk factors, and thus reliable support services play a crucial role in post-disaster operations and the computation of damage/loss. The following is an economic analysis, involving an estimate of the costs when a structure collapses:

$$C = r + g \times \frac{P}{l}$$

where C $\Rightarrow$ the cost; r $\Rightarrow$ a multiple of the baseline cost; g $\Rightarrow$ a multiple of the initial baseline cost; P $\Rightarrow$ the notional risk failure; l $\Rightarrow$ the discount rate on economic valuation.

We may derive another useful expression for reliability, which is the probability that a system performs a specific function or mission under given conditions for a prescribed time. Let R(t) be the probability that it will fail during the time period; (t) the hazard rate; F(t) the cumulative failure probability; f(t) the failure probability density.

The relevant equations are:

(1) $f(t)=\frac{dR(t)}{dt}$

(2) $(t)=\frac{d\ln R(t)}{dt}=\frac{1}{R(t)}\times\frac{dR(t)}{dt}=\frac{f(t)}{R(t)}$

(3) $F(t)=1-R(t)$

Integrating (1) and (2), we have:

(4) $R(t)=\exp\left[-\int_0^t \lambda(t')dt'\right]$

(5) $f(t)=\lambda(t)\exp\left[-\int_0^t \lambda(t')dt'\right]$

Incidentally, Mean Time to Failure (MTTF) of a component of the system is given by:

MTTF = $\int_0^\infty R(t)dt$, and in particular for random failures, MTTF = $\frac{1}{\lambda}$

Dr Gerhard Berz of Munich Re studied the cost of disasters and its implications for the insurance industry. Dr Berz defines ‘insurance’ as a particular form of disaster preparedness; hence the latter is a prerequisite for many economic activities. Earthquake risk may be defined as the probability percentage of the occurrence of a continental earthquake resulting in ground accelerations greater than 0.1g during a specific period, say ‘p’ years. The seismic risk or R(p) is expressed as follows:

$$R(p)=1-\exp[1-xp(-UM_s)]$$

where x $\Rightarrow$ mean rate of occurrence per year; U $\Rightarrow$ the slope; M_s $\Rightarrow$ the surface magnitude wave.

Probabilistic analysis is the usual starting point for risk assessment. Indeed, there is no single or simple/reductive mathematical formula for tracing the complex relation between the cause and the quantitative ‘effect’ of natural disasters. Mathematical and deterministic models may have to be combined with an empirical approach and sometimes even with an intuitive knowledge of the damage inflicted.

Select Bibliography

Bazerman, Max H, and Watkins, Michael B, *Predictable Surprises: The Disasters You Should Have Seen Coming and How to Prevent Them*, Harvard Business School Press, Cambridge MA, 2004.

Comfort, L K, ed, *Managing Disaster: Strategies and Policy Perspectives*, Duke University Press, Durham, 1988.

Drabeck, T E, *Human System Responses to Disasters: An Inventory of Sociological Findings*, Springer Verlag, New York, 1986.

Davis, Ian, *Shelter After Disaster*, Oxford Polytechnic Press, Oxford, 1978.

Fisher III, H W, *Response to Disaster*, University Press of America, Maryland, 1994.

Friedman, Thomas L, *The World is Flat: A Brief History of the Globalized World in the Twenty-first Century*, Allen Lane, London, 2005.

Gupta, Harsh K, *Disaster Management*, Universities Press (India) Pvt Ltd, Hyderabad, 2003.

Hewitt, K, *Interpretation of Calamity from the Viewpoint of Human Ecology*, Allen and Unwin, London, 1983.

Holt, Barry, *True Risk Management, Preventing Natural Disasters: The Role of Risk Control and Insurance*, The Royal Society, UK General Accidents, 1995.

Leet, L Don, *Causes of Catastrophe: Earthquakes, Volcanoes, Tidal Waves and Hurricanes*, McGraw Hill, London and New York, 1948.

Lighthill, James et al, ed, *Tropical Cyclone Disasters*, Peking University Press, Beijing, 1992.

Liverman, Diana M, 'Vulnerability to Global Environmental Change' in Cutter, Susan L, ed, *Environmental Risks and Hazards*, Prentice Hall India, New Delhi, 1999.

Maskrey, Andrew, *Disaster Mitigation: A Community Based Approach*, Oxfam, Oxford, 1989.

Mitchell, James K, 'Confronting Natural Disasters: An International Decade for Natural Hazard Reduction', *Environment*, 1988, Vol. 30, No. 2, pp.25-9.

— 'Where Might the Decade for Natural Disaster Reduction Concentrate its Activities?', *Technical Report*, The Natural Hazard Research and Application Information Centre, Boulder, Colorado, 1988.

Murty, T S, 'List of Major Natural Disasters (1960-1987)', *Natural Hazards*, pp.303-4.

WMO, *Natural Disaster Reduction*, 1989.

Pan American Health Organization (PAHO), *Natural Disasters: Protecting the Public Health*, WHO, Washington DC, 2002, p.575.

Pelling, Mark, ed, *Natural Disasters and Development in a Globalizing World*, Routledge, London, 2003.

Perez de Cuellar, Javier, 'IDNDR: A Challenge for the United Nations', UNDRO News, Jan/Feb 1990, pp.2-3.

Press, Frank, *Report on the International Decade for Natural Disaster Reduction,* General Assembly, Economic and Social Council, UN, 1989.

— 'The Role of Science and Technology in the IDNDR', UNDRO News, May/June 1990, pp.1-13.

— *The Need for Action, Disaster Reduction, Coping with Natural Hazards,* UNESCO, 1993.

Quarentelli, E L, 'Converting Disaster Scholarship into Effective Planning and Managing: Possibilities and Limitations', *International Journal of Mass Emergencies and Disasters,* March 1993, pp.15-38.

Sanderson, David et al, *Disaster Mitigation, Preparedness and Response: An Audit of UK Assets,* IT Publications, Oxford Centre for Disaster Studies, 1996.

Sharma, Vinod K, *Disaster Management,* Indian Institute of Public Administration, New Delhi, 1994.

Sinha, D K, ed, *Readings on Natural Disaster Reduction and Media,* Indian Science News Association, Calcutta, 1992.

— 'Disaster Studies: In Retrospect', *Journal of the Department of Environment and Forests, Govt of West Bengal,* Banabithi, Calcutta, 1990.

— *Some Remarks on Management of Disasters Pertaining to Coastal Areas in the Coastal Management of West Bengal,* Sea Explorer's Institute, Calcutta, 1990.

— 'Pre-disaster Warnings', *The Telegraph,* Special Supplement, Calcutta, 31 October 1990.

— 'International Decade for Natural Disaster Reduction (IDNDR): A Critique on Information Management', *CSI Silver Jubilee Convention Souvenir,* Calcutta, 1990, pp.47-9.

— 'On Climatic Change and Natural Disasters', *Science and Culture,* Indian Science News Association, Calcutta, 1992, Volume 58, pp.106-11.

— 'Coping with Natural Disasters: An Integrated Approach', *The Shaping of Indian Science,* Indian Science Congress Association, Universities Press (Pvt) Ltd, Hyderabad, 2003, Volume 3, pp.1725-71.

— 'Natural Disaster Reduction for Nineties Perspectives, Aspects and Strategies', International Journal Services, Calcutta, 1990.

— *Towards Emerging Facets of Disaster, Emergency and Crisis Management,* Research Book Centre, New Delhi, 2006.

Sinha, D K, Sen, G K, and Chatterjee, M, *Dynamics of the Lower Troposphere: A Diagnostic Critique of Pre-Monsoon and Monsoon Scenarios,* University of Calcutta, Calcutta, 1996.

Tierney, Kathleen J, Lindell, Michael K and Perry, Ronald W, *Facing the Unexpected: Disaster Preparedness and Response in the United States,* Joseph Henry Press, Washington DC, 2001.

Timmerman, P, *Vulnerability, Resilience and the Collapse of Society,* Environmental Monograph No.1, Institute of Environmental Studies, University of Toronto, 1981.

Wildavsky, Aaron and Dake, Karl, 'Theorizing Risk Perception: Who Fears What and Why?' in Cutter, Susan L, ed, *Environmental Risks and Hazards,* Prentice Hall India, New Delhi, 1999.

